Shared Lives of Humans and Animals

Animals are conscious beings that form their own perspective regarding the lifeworlds in which they exist, and according to which they act in relation to their species and other animals. In recent decades a thorough transformation in societal research has taken place, as many groups that were previously perceived as being passive or subjugated objects have become active subjects. This fundamental reassessment, first promoted by feminist and radical studies, has subsequently been followed by spatial and material turns that have brought non-human agency to the fore. In human–animal relations, despite a power imbalance, animals are not mere objects but act as agents. They shape our material world and our encounters with them influence the way we think about the world and ourselves.

This book focuses on animal agency and interactions between humans and animals. It explores the reciprocity of human–animal relations and the capacity of animals to act and shape human societies. The chapters draw on examples from the Global North to explore how human life in modernity has been and is shaped by the sentience, autonomy, and physicality of various animals, particularly in landscapes where communities and wild animals exist in close proximity. It offers a timely contribution to animal studies, environmental geography, environmental history, and social science and humanities studies of the environment more broadly.

Tuomas Räsänen is a postdoctoral researcher at the Department of European and World History at the University of Turku, Finland. His research interests include animal history, environmental history and history of science. He has published several articles concerning the history of marine sciences and politics in the Baltic Sea area and the history of Finnish environmentalism.

Taina Syrjämaa is Professor of European and World History at the University of Turku, Finland. She has studied urban history, the history of the exhibition medium, the belief in progress, historical spatiality and diffuse agency. She is currently leading the research project "Animal Agency in Human Society."

Routledge Human–Animal Studies Series

Series edited by Henry Buller

Professor of Geography, University of Exeter, UK

The new *Routledge Human–Animal Studies Series* offers a much-needed forum for original, innovative and cutting-edge research and analysis to explore human–animal relations across the social sciences and humanities. Titles within the series are empirically and/or theoretically informed and explore a range of dynamic, captivating and highly relevant topics, drawing across the humanities and social sciences in an avowedly interdisciplinary perspective. This series will encourage new theoretical perspectives and highlight ground-breaking research that reflects the dynamism and vibrancy of current animal studies. The series is aimed at upper-level undergraduates, researchers and research students as well as academics and policy-makers across a wide range of social science and humanities disciplines.

Critical Animal Geographies: Politics, Intersections and Hierarchies in a Multispecies World
Edited by Kathryn Gillespie and Rosemary-Claire Collard

Urban Animals: Crowding in Zoocities
Tora Holmberg

Affect, Space and Animals
Edited by Jopi Nyman and Nora Schuurman

Animal Housing and Human–Animal Relations: Politics, Practices and Infrastructures
Edited by Kristian Bjørkdahl and Tone Druglitrø

Shared Lives of Humans and Animals: Animal Agency in the Global North
Edited by Tuomas Räsänen and Taina Syrjämaa

Shared Lives of Humans and Animals

Animal Agency in the Global North

Edited by Tuomas Räsänen and Taina Syrjämaa

Routledge
Taylor & Francis Group
LONDON AND NEW YORK

First published 2017 by Routledge

2 Park Square, Milton Park, Abingdon, Oxfordshire OX14 4RN
52 Vanderbilt Avenue, New York, NY 10017

Routledge is an imprint of the Taylor & Francis Group, an informa business

First issued in paperback 2018

British Library Cataloguing in Publication Data
A catalogue record for this book is available from the British Library

Library of Congress Cataloging in Publication Data
A catalog record has been requested for this book

ISBN: 978-0-415-41925-3 (hbk)
ISBN: 978-0-367-21886-7 (pbk)

Typeset in Times New Roman
by diacriTech, Chennai

Contents

Illustrations

Figures

Tables

Contributors

Vernae Angnaboogok is from Wales, Alaska, traditionally known as Kiŋigin, an Iñupiaq village located at the northwestern most tip of the Seward Peninsula. Her family instilled in her the passion for learning and embracing the Iñupiaq way of life. She contributed to the research as an independent researcher from the University of Alaska Fairbanks Rural Development program and as an intern with the Kawerak Social Science Program.

Outi Autti is a postdoctoral researcher at the Arctic University of Norway in Tromsø (UiT). Previously she has worked in various positions at the University of Oulu, Finland. Her research fields are environmental sociology, rural sociology, cultural studies, and human geography. She studies the social dimensions of environmental and cultural change in northern communities.

Pia Bäcklund is a postdoctoral researcher at the University of Tampere, School of Management. Her research themes have addressed a wide scope of issues concerning planning theories, democracy, participation, local development projects, knowledge management, and political subjectivity of animals.

Karin Dirke is an associate professor/senior lecturer in History of Ideas at the Department of Culture and Aesthetics, Stockholm University. Her research mainly concerns human–animal studies and she has published several works on the historical relationships between humans and other animals, both wild and domestic.

J. Dwight Hines is an Associate Professor of Global Cultural Studies in the Humanities and Human Sciences Department of Point Park University in Pittsburgh, PA. His primary research interests revolve around the human and ecological consequences of changes facing the Greater Yellowstone Ecosystem in the northern (US) Rocky Mountains, especially those changes precipitated by the ongoing gentrification of the rural and small-town spaces therein and the rising influence of the post-industrial "experience economy."

Stella Hockenhull is a Reader in Film and Television Studies and Co-Director of the Research Centre Film, Media, Discourse and Culture at the University of

Wolverhampton. She has built up a strong research profile in British cinema and landscape and published widely on this. Her more recent work focuses on horses in film.

Linda Kalof is Professor of Sociology, Community Sustainability, Animal Studies and Environmental Science & Policy at Michigan State University. Her current research focuses on the illustration of animals in popular science and is funded by the National Science Foundation.

Leena Koski is a Professor of Sociology at the Department of Social Sciences, University of Eastern Finland. She has addressed widely the questions considering the historical construction of moral orders in elementary, adult, and university education, gender in education, and the social and symbolic spaces for companion animals.

Riitta-Marja Leinonen is a University Teacher of Cultural Anthropology at the University of Oulu in Finland, where she also received her PhD in cultural anthropology. She did her doctorate on the meaning of the horse in Finland in the twentieth century using ethnographic and oral history methods. She has also studied the human–horse relationship at war through the narratives of Finnish WWII veterans, the practices and conceptions of the death of a horse, and nationalistic representations of the Finnhorse in trotting championships.

John Martin is Professor of Agrarian History at De Montfort University Leicester. In addition to his books and articles, he has acted as a consultant and contributor to a number of television and radio series concerning the transformation of British agriculture.

Jukka Nyyssönen researcher, has worked in different positions in the departments of history at the University of Jyväskylä (Finland) and University of Tromsø – The Arctic University of Norway (UiT). Currently he is affiliated to the Department of Cultural Sciences at the UiT and Section for Cultural Studies at the Tromsø University Museum. Nyyssönen has published in the fields of history of minorities, environmental history, and history of science, with focus on, for example, Sami identity politics, Sami historiography, and research on Skolt Sami.

Julie Raymond-Yakoubian is an anthropologist and the Social Science Program Director at Kawerak, Incorporated. For the past ten years she has carried out collaborative research with Tribes in the Bering Strait Region of Alaska through community-based research projects. Her recent work has included research on traditional knowledge of ice seals, walruses, fisheries, ocean currents, climate change, and the supernatural environment.

Harriet Ritvo is the Arthur J. Conner Professor of History at the Massachusetts Institute of Technology. She is the author of several books and numerous articles on British cultural history, environmental history, the history of human–animal relations, and the history of natural history. Her current research concerns wildness and domestication.

Jessica Bell Rizzolo is a PhD student in Sociology at Michigan State University, with specializations in animal studies, environmental science and policy, and conservation criminology, and Director of the Asian Elephant Program at the Kerulos Center. Her research areas include trans-species psychology, discursive representations of wildlife, the sociopolitical dynamics of conservation initiatives, wildlife tourism, and the illegal wildlife trade.

Tuomas Räsänen is a postdoctoral researcher at the Department of European and World History at the University of Turku, Finland. His research interests include animal history, environmental history and history of science. He has published several articles concerning the history of marine sciences and politics in the Baltic Sea area and the history of Finnish environmentalism.

Nora Schuurman is an Adjunct Professor in Animal Geography at the University of Eastern Finland, Joensuu campus. Her research interests include human–animal relationships, animal-related practices, emotions, and expertise, and the role of the horses and pets in Western society. Her current research focuses on the conceptions, experiences, and practices concerning companion animal death.

Kirsi Sonck-Rautio is a cultural anthropologist and a doctoral candidate at the University of Turku, Finland. She obtained her MA from the University of Oulu, Finland. She is currently conducting fieldwork for her PhD dissertation in the archipelago of southwest Finland and has studied local small-scale fisheries for years.

Taina Syrjämaa is Professor of European and World History at the University of Turku, Finland. She has studied urban history, the history of the exhibition medium, the belief in progress, historical spatiality and diffuse agency. She is currently leading the research project "Animal Agency in Human Society."

Stephen Vrla is a dual PhD student in Sociology and Curriculum, Instruction, and Teacher Education at Michigan State University. He is earning specializations in animal studies and environmental science and policy. His research focuses on the intersection of environmental education with deliberative democracy. Stephen is also a naturalist and youth leadership program coordinator at Fenner Nature Center in Lansing, Michigan.

Cameron Thomas Whitley is a PhD student in Sociology at Michigan State University, with specializations in animal studies, environmental science and policy, and gender justice and environmental change. Cameron's current research explores the hidden place of animals in energy development.

Acknowledgements

The book project was launched in 2014 in Turku with the Urban–Rural–Wilderness workshop, where almost all the authors were present to discuss the main themes of this book and to present the first versions of their respective chapters. This book was made possible by the research project Animal Agency in Human Society (274573), led by Taina Syrjämaa and funded by the Academy of Finland.

We would like to thank Arthur J. Conner Professor Harriet Ritvo for her generous support throughout this book project during which we were honored to have her twice as our guest in Turku. She also hosted one of the editors of this book while he worked as a visiting scholar at MIT in the spring of 2016.

Introduction

Tuomas Räsänen and Taina Syrjämaa

The experience of modernity in the industrial North has led to radical changes in human–animal relations, which has evolved into an ever more complex dynamic. On the one hand, humans have distanced themselves from animals. The majority of consumers eat meat reared in factory farms and the "wild" animals we encounter are often actually carefully controlled by experts using modern technological devices in confined "natural" environments. On the other hand, humans in the affluent North have developed an effective relationship with animals that was rarely seen in pre-modern societies. Hundreds of millions of people share their lives with pets, which have been turned into co-consumers, while the welfare of animals has attained unprecedented political and social appeal. All these processes derive from historical and cultural developments, which are the focus of this book.

In the past decade the realization that human welfare depends on animal lives has led many scholars in the humanities and social sciences to examine human–animal relations. One of the most widely adopted approaches has been to study human images of and attitudes to animals. Another popular approach concerns animal ethics, in which scholars have studied the human use and abuse of animals and the possibility of animal rights. Most recently, a third approach has begun to focus on animal agency.[1] We adopt this last stance and aim to bring to the fore animal species and individual animals as actors in human communities.

Animals are conscious beings that form their own perspective regarding the life-worlds in which they exist, and according to which they act in relation to their species and other animals. In human–animal relations, despite a power imbalance, animals are not mere objects but act as agents. Animals shape our material world and our encounters with them influence the way we think about the world and ourselves.

In recent decades a thorough transformation in societal research has taken place, as many groups that were previously perceived as being passive or subjugated objects have become active subjects. This fundamental reassessment, which was first promoted by feminist and radical studies, has subsequently been followed by spatial and material turns that have brought non-human agency to the fore. Human–animal studies belong to the same continuum, in the sense that they examine a previously ignored group of actors and foreground non-human agency.

The essential challenge of human–animal studies is to examine animal "others" that have left no direct imprint in source material. Animals can only be understood through the lens of human perception and comprehension, whether based on scientific, religious, affectionate or other premises. It is impossible to evade human mediation, as we do not know what animals think. Yet, animals are very real and are not mere reflections of the human imagination. The historian and historical ethologist Éric Baratay has pointed out that ethnological and historical research are always essentially examinations of an "other." The "other" is usually another human from a different cultural and social setting, or in the case of history, even from another era. With this in mind, the study of animals can be viewed as being a similar research undertaking. However, what is different here is the degree of otherness, rather than the epistemological constellation.[2]

The amount of literature on human–animal relations has greatly expanded in recent years. Contemporaneously, the notion of animal agency has gradually started to appear in scholarly works more frequently. However, the content of this concept has often remained vague and obscure. This volume brings together fourteen chapters, which all combine theoretical and empirical inquiry, in a joint effort to make animal agency visible and analyzable. Our aim is not to seek a fixed criterion for the concept or a settled analytical method. Instead, the authors represent different disciplines and divergent theoretical points of departure regarding animal agency: from an all-inclusive Latourian concept of agency to a school of thought that sets higher cognitive barriers, such as the ability to communicate and make intentional decisions, for an animal as an agent. By so doing, the book explores multiple aspects of animals' interactions with humans and offers multifaceted perspectives on the topical discussion of animal agency.[3]

While issues pertaining to human–animal relations have global dimensions, they are also deeply connected to environmental, cultural and historical contexts. This book strongly underlines the essential importance of environmental, social and temporal contextualization in the process of coming to terms with animal agency. Our chosen focus is on the American and European North since the nineteenth century, which is characterized by tension and contradiction: industrialization and urbanization in these regions brought about radical changes in the human–animal relationship, yet these zones continue to be marked by areas of wilderness and sparsely populated areas of countryside. Focusing on these regions makes it possible to examine a large area, but contemporaneously to take into account national, regional and temporal differences.

In this book, nineteen scholars, representing various strands of humanistic and societal research, offer empirical case studies on how animal agency has had – and continues to have – a profound influence on individuals and entire societies. The human–animal relations examined in this volume not only cover different locations, but also a wide variety of animals – both wild and domesticated – that interact with humans. The chapters in this book also scrutinize co-living in a variety of different environments, including cities, rural regions and wilderness areas.

The book is divided into three parts, which each emphasize the different ways in which animal agency is manifested in human–animal relations. In the first section, the focus is on individuals in interspecies interactions. The human propensity to perceive animals as an anonymous mass and (mis)treat, ignore and/or deny their individuality accordingly is deeply rooted in our culture. This tendency has sustained a fundamental belief in the imaginary otherness of animals and has endorsed their disempowerment. The degree of otherness dissipates, however, when the individuality of animals is acknowledged and the focus of research is placed on individuals from different species and their respective encounters.

In the first chapter, the sociologists Leena Koski and Pia Bäcklund discuss canine agency in the popular context of contemporary dog sports. They show how humans set boundaries for the conduct of dogs, but that ultimately dog agility is based on the activity of both a dog and a trainer as individuals. Nora Schuurman continues this discussion by focusing on tacit knowledge as an essential element of contemporary equestrian culture. She examines how horses are acknowledged agents that are able to make and express their choices. These chapters are based on oral or written accounts which were created by people who experienced continuous interactions with animals and thereby gained practical skills in interspecies communication.

Art offers another manner of exploring animal viewpoints. In her chapter, Stella Hockenhull explores the age-old human–horse relationship in Iceland by analyzing a film that privileges the equine perspective. The narration of this non-anthropomorphizing film connects the stories of individual horses and men and unites them to the landscape of the Icelandic wilderness, which forms an integral part of interspecies interaction. In her ethnological study, Riitta-Marja Leinonen traces practices of human–horse co-living in rural folklore, which emphasizes mutuality in the sharing of everyday life and space. The first part of the book closes with Taina Syrjämaa's study on early dog shows. She shows how these events, which manifestly celebrated breeding and human attempts to manipulate animals, actually turned out to be meeting points for individuals, both human and canine, who affected each other's lives.

In the second section of this volume the focus turns from individual animals to populations. More specifically, we examine how humans and animal populations have formed interspecies communities, in which a group of animals display agency by affecting the whole community, including human societies. These interspecies communities are susceptible to conflict, but also to cooperation and mutual learning.

In his chapter, John Martin examines the long-term relationship between humans and rabbits in late nineteenth- and early twentieth-century Britain, in which rabbits gradually lost their esteemed status as a game species, due to a conflict over agricultural production. The impact of the rabbit population on agricultural output in the first part of the twentieth century threatened the economy as a whole and compelled British society to develop a radical means to eradicate rabbits.

In interspecies communities certain animals often act as intermediary agents that convey knowledge to humans pertaining to changes in the environment. Consequently, this makes such animals a powerful social force. Tuomas Räsänen shows how the declining white-tailed eagle population in the northern Baltic Sea area acted as a warning signal for humans, who subsequently enacted laws to inhibit the contamination of the environment. By so doing, the interaction between humans and eagles instigated changes in environmental thinking and benefited humans and eagles alike.

Julie Raymond-Yakoubian and Vernae Angnaboogok examine the changing relationship between Bering Strait Eskimos and fish. The indigenous people of this region traditionally perceived fish as being members of their society. According to this perspective, fish possessed almost human-like agency, with intelligence and personhood. Consequently, humans had to negotiate with them in order to secure catches. However, with the encroachment of Christian beliefs and a capitalist economy, fish gradually began to be seen as being devoid of agential capabilities and were conceptualized as a commodifiable natural resource.

In her chapter, Kirsi Sonck-Rautio studies the transformation of a local winter seine fishing community in the Baltic Sea region. This traditional subsistence-based community relied heavily on fish. Utilizing a socio-ecological system approach, she analyzes how communities adapted to the movements and choices of Baltic herring shoals, as well as icy living conditions and global changes.

The third part of this volume addresses how animal agency has been interpreted and represented in social and political terms, including oral traditions, literature and the mass media. Agency may be a novel concept in scholarly discussions of animals, but, as textual representations of past human–animal relations demonstrate, humans who have been in regular contact with animals have seen animals as agential beings. Representations are powerful tools in the construction of the human–animal relationship, irrespective of whether they are scientific, fictional, written or oral. Consequently, as Karin Dirke argues in her chapter in this book, this perceived animal agency "prompts humans into … behavioral performances."

In his contribution to this volume, Jukka Nyyssönen examines the agency of reindeer in three texts from different eras, which all depict the animals as agents that cope with human presence in their lives according to their own (human-controlled) preferences. Nyyssönen shows how reindeer act as trafficking agents between nature and culture, shifting between unpredictable behavior that is independent of human agency and predictable aspects of domesticity. The latter feature is harnessed for human use. However, he also argues that reindeer agency, as represented in the texts analyzed by the author, has significantly changed during the course of the modern era. Thus, Linnaeus's sense of awe when observing reindeer agency gave way to more instrumental conceptions of the species in the twentieth century.

In Outi Autti's chapter, we find the same trafficking of animal agency between nature and culture in oral representations of the inhabitants of riverside communities in northern Finland. People's reminiscences about salmon, for example, describe how they used to climb up rivers before the construction of hydroelectric plants. In a similar manner to the role of reindeer for the Sami, salmon had identificational significance for the riverside communities, according to which they projected their own life situation. The agency of salmon helped them to navigate in the world.

Linda Kalof, Cameron Whitley, Stephen Vrla and Jessica Bell Rizzolo identify the use of anthropogenic food sources as a focal point of multi-species communities. In their study, the authors examine visual representations of liminal animals, as reproduced in *National Geographic* since the 1880s. The authors argue that human–liminal animal contact has been – and will increasingly be – essential in the formation of our understanding of ecology.

In the penultimate chapter of our volume, Karin Dirke examines how narratives concerning the encounters of humans and large carnivores in Sweden changed during the course of the nineteenth century. She argues that as this century progressed, large carnivores were ascribed more agency. At the same time, however, the agency of wolves and wolverines, in particular, began to be portrayed in more threatening terms. Yet, the emphatic adoption of the carnivore perspective in hunting literature and newspapers provided a forum for advocating the protection of the Brown Bear and the Lynx.

In the final chapter, J. Dwight Hines analyzes how the agency of wolves living in Yellowstone National Park changed human actions and values in the United States in the twentieth century. He argues that wolves have been a symbol of land-use conflicts in the American West, in which others see public land as primarily being a place of production, while the post-industrial middle class, in particular, emphasized the production of experiences. The latter viewpoint is reinforced by the conservation ethos surrounding the wolf, in which the species is perceived as being an ecologically important apex predator.

The volume closes with an epilogue, in which renowned animal historian Harriet Ritvo provides concluding remarks about the discussion on animal agency.

***

Human–animal relations is one of the most pressing problems of the twenty-first century. Currently a quarter of all mammal species, for example, are considered endangered by the International Union for the Conservation of Nature. The number of extinct species is growing every year. According to a report by the World Wildlife Fund, the total number of terrestrial and aquatic animals has fallen by approximately fifty percent during the past forty years. This loss of wildlife has been substituted by a massive growth in the number of domesticated animals. The global pig population currently stands at approximately one billion, while there are also one and a half billion cows in the world. The sole purpose of these vast populations is to feed humanity. Every year a staggering 60 billion chickens are slaughtered for food.[4] The consumption of meat is projected to grow rapidly in the developing world. All of these trends are not only ethically dubious, but are also having a devastating effect on global ecology, climate change and the quality of soil and water.

However, it is worth remembering that, while the affluent countries in North America and Europe eagerly present themselves as champions of the responsible use of the natural world, the loss of biodiversity and the exploitation of animal lives has been pushed to the extreme particularly in the global North. The consumption of animal products in North America and northern Europe surpasses anything seen in human history; the rapid destruction of wildlife, which the developing world is facing today, has been going on in the global North for centuries.

In recent decades, numerous animal rights and environmental organizations, particularly in the global North, have tried to advance the cause of animals, but results have been limited. In recent years, improvements have been made through legislation and international treaties to the practices of treating animals and a growing number of people have decided to refrain from consuming animal products. However, progress has been only local and sporadic when looking at the larger picture. Sue Donaldson and Will Kymlicka, among others, have pointed to the undeniable failure of the animal rights movement. They argue for a new animal ethics, whereby animals would be entitled to the right of citizenship alongside humans.[5] This may well be a valuable step in the right direction, although we are aware that an ethical construction will inevitably fail to alter human behavior if it is not accompanied by an understanding of the role of animals in human culture. Animals are not just bystanders in a community, but are also active participants who earn their citizenship. In this book, our aim is to produce an awareness of the multiple ways animals form a fundamental part of our everyday lives and experiences, without which the development and functioning of human society would be impossible. We are indebted to animals as much as to our own ingenuity and industriousness for our present well-being. The least we can do in return is to recognize how animals merit a deserved place in the analysis of past and present human societies, and treat them with more respect than has hitherto been the case.

Notes

1 See Freeman, Leane and Watt, *Considering Animals;* Nance, *Entertaining Elephants;* McFarland and Hediger, *Animals and Agency*.
2 Baratay, *Le point de vue animal*.
3 Similarly to the divergent conceptions of agency, the authors refer to animals in different ways. For example, some authors use "which," some "who." Others refer to animals as "she" irrespective of the sex of the animal individual, while others distinguish between males and females.
4 WWF, "Half of World Wild Life Lost"; IUCN, "IUCN Red List Factsheet"; FAO, "FAO Statistical Pocket Book 2015"; The Poultry Site, "Global Poultry Trends 2014."
5 Donaldson and Kymlicka, *Zoopolis*, 4–14.

Bibliography

Baratay, Éric. *Le point de vue animal. Une autre version de l'histoire*. Paris: Seuil, 2012.
Donaldson, Sue, and Will Kymlicka. *Zoopolis: A Political Theory of Animal Rights*. Oxford, UK: Oxford University Press, 2011.

FAO. "FAO Statistical Pocket Book 2015, World Food and Agriculture." Accessed March 30, 2016. www.fao.org/3/a-i4691e.pdf.

Freeman, Carol, Elizabeth Leane, and Yvette Watt, eds. *Considering Animals: Contemporary Studies in Human-Animal Relations*. Farnham, UK: Ashgate, 2011.

IUCN. "The IUCN Red List of Threatened Species." Accessed February 7, 2017. www.iucn.org/about/work/programmes/species/our_work/the_iucn_red_list/.

McFarland, Sarah E., and Ryan Hediger, eds. *Animals and Agency: An Interdisciplinary Exploration*. Leiden, Netherlands: Brill, 2009.

Nance, Susan. *Entertaining Elephants: Animal Agency and the Business of the American Circus*. Baltimore, MD: Johns Hopkins University Press, 2013.

The Poultry Site. "Global Poultry Trends 2014." Accessed March 30, 2016. www.thepoultrysite.com/articles/3230/global-poultry-trends-2014-poultry-set-to-become-no1-meat-in-asia/.

WWF. "Half of Global Wildlife Lost, Says New WWF Report." Accessed March 30, 2016. www.worldwildlife.org/press-releases/half-of-global-wildlife-lost-says-new-wwf-report.

Part I

Co-Living Individuals

1 Whose Agency?

Humans and Dogs in Training

Leena Koski and Pia Bäcklund

Introduction

If left to their own devices, dogs as companion animals living in human homes are not very tolerable. Their activities, behaviors, habits, and ways of communicating are those of animals; therefore, it is generally thought that their characters must be shaped and trained. The boundaries of a modern, urban dog's controlled space and activities are quite strict; dogs are accepted into modern human communities in subordinate roles, submissive to human rules and control. Simultaneously, dogs are often expected to play an active role in their keepers' lives. They are taken into human homes for some more or less specific purposes, such as "versatile companionships,"[1] and are expected to develop individual personalities similar to those of other family members.[2]

One of the fairly new roles that dogs have occupied is as companions in various dog sports. In this context, dogs are expected to act as active partners in learning and achieving the human's chosen goals. Goal-oriented training requires the development of effective interspecies communication and cooperation between two partners in order to achieve any success. Cooperation entails the idea that the dog is capable of certain levels of cooperative communication, and is therefore in possession of some kind of agency. Our questions are therefore as follows: (1) How is the dog's agency constructed in the cooperation process? (2) How is the dog herself displaying her agency? (3) What does interspecies cooperation mean in relation to the respective agencies of both the human and the dog? The space for a discussion on the agency of dogs is comprised here of three defining structures. First, there is a presentation of the general and historical conditions defining the lives of dogs in Finland. Second, we discuss the intensified interactions between the dog and the human, and third, we explain the simultaneous cultural discourses of humanizing, instrumentalizing, and animalizing dogs in modern societies.[3]

Finland was urbanized relatively late, in just the past few decades. Thus, traces of a more traditional understanding of the dogs' social place and status as living in the countryside hunting elk, bear, squirrel, or fowl as well as managing themselves outdoors during the cold winters and guarding the barnyard are still present. The dogs as family members sharing human homes, and as present in urban, public places, are relatively new phenomena as compared to their reception in most other European cultures, which have many more years of urbanized traditions. The late urbanization in Finland implies a shorter developmental period for urban mentality as well. Therefore, dogs being visibly present in the majority of Finnish urban environments is fairly novel. Furthermore, there are tensions and conflicts leading to the demand for increasingly strict control over public places that either deny or allow dogs; dogs are not, for example, allowed onto public beaches, playgrounds, or ski tracks, inside shops or offices, or at the marketplaces. They are only permitted off-leash within fenced areas or with the landowner's permission, and only recently have they been allowed inside restaurants with express management permission. However, the demands for more control in Finland have not yet reached the level of defining any particular breeds as more dangerous than others, as is the case in Norway, for example.[4] Our starting point, then, is that late urbanization affects the way in which we see dogs as companion animals in general, which also affects how their agency is defined in dog sports, and what these dogs now appear to need for a good life.

Training dogs to perform highly-specified tasks is an exemplary case where the possible and actual interpretations, constructions, and representations of a dog's agency can be traced. Training is an intensified case of interaction and interspecies cooperation. Training for a fixed goal, such as participation in official competitions with the expectation of performing certain exercises, exposes both the trainers' and the dogs' subtle forms of interaction and learning.[5] Power notes that inside homes, as opposed to at competitions or other spaces outside of the home, "humans and dogs experienced a greater freedom to define their own relations and presences."[6] Thus, our data refer to the public sphere, where the agency or agencies are presupposed or given to the dog, and where the preconditions for the dog to display her agency are already presented.

Dogs and Agency: A Theoretical Background

Agency is one of the most troublesome concepts in the social sciences; it demarcates a contradictory space, including questions of self-fulfillment, forms of interaction, political possibilities, and restrictions in structural and everyday power relations. In the case of the dog, such a concept becomes even more difficult to define. The social space for dogs is both inside and outside of human society[7] and we have a vague (if any) understanding of the dog's consciousness and ideas of self-fulfillment. However, as Bekoff notes, when discussing the possibility of morality with respect to animal behavior, play and cooperation are the keys of understanding and the grounds of establishing such moral rules and behavior. Play cannot occur without rules; it requires fairness and trust on both parts. If the rules are broken, playing stops and continues only if the playmate "apologizes" to the other, who then decides whether to continue playing or not.[8] Based on Bekoff's ideas, McFarland and Hediger claim that play and "its underlying rules reflect animal agency," because play requires that participants will "follow the rules (which means that they also could choose *not* to)."[9]

Furthermore, Bekoff highlights the importance of cooperation over competition; fairness and trust are prerequisites for play, and cooperation is a more successful resource for survival than competition or "survival of the fittest," at least for pack animals such as dogs. Thus, play requires a kind of cooperation that drives behavior in a more productive direction than competition does.

Drawing again on Bekoff's arguments McFarland and Hediger claim that play is the central element and common denominator of animal agency, since it "requires a choice, and choice is part of what defines agency."[10] Bekoff's study is based on observations of domestic dogs and their wild relatives, such as coyotes and wolves, and this explains why these ideas of a basic understanding of canine agency, inscribed with its own kind of morality, are relevant as theoretical starting points for our study. As we have shown in our earlier studies, cooperation, fair play, and an emphasis on the dog's own will in the training processes are at the very core of the hegemonic pedagogical ideas of dog training in Finland.[11] The pedagogy includes the idea that the dog displays her individual feelings and affectations during the training process, and that these factors must be taken into account. Thus, dogs are supposed to make choices, and these choices are indicators of their agency.

However, our data does not consist of observations of canine behavior, but instead includes the dog trainers' descriptions of their training methods, how their dogs learn, the challenges faced while training and in everyday life, the demonstrations of educability on behalf of their canine partners, and descriptions of the ideal dog. Theoretically and methodologically, we are discussing the human interpretation of an agency "given" to, or "taken" from dogs. Therefore, we believe that discussing only how agency can be understood among the dogs themselves is not enough. What must also be considered is how dog agency is defined in relation to human agency in order to form a more comprehensive understanding of these agencies in their interspecies cooperation. Here, we turn to feminist discussions for more illumination of agency-related issues.[12]

In feminist discussions, agency is understood as dynamic and multifaceted; it is intertwined in historical and cultural conditions and power relations, and it is discursive, material, and situational. As a situational concept, agency varies depending upon the conditions of time, space, materiality, bodies, and relations to other agents and their positions.[13] It can be suggested that when humans are interacting with dogs, the formation of human agency in this particular context is situational, related to that of the dog and vice versa in an interspecies collective. Both the trainers and the dogs have to adjust their actions to those of their partner, making situational choices that change each other's behaviors.

Sanders identifies numerous stories describing how dogs "acted in ways that were thoughtfully intended to shape their owner's definitions of the situation and to manipulate their subsequent behavior to desirable ends."[14] A dog "shaping" her owner's behavior is a definition of agency. The dogs' agency is related to the fact that dogs have a hybrid existence in the sense in which Latour defines hybridism; dogs are creatures that would not even exist as a species without human investment,[15] and, as Haraway suggests, humans would not exist as they are without their historical relations with dogs: "Dogs are about the inescapable, contradictory story of relationships – coconstitutive relationships in which none of the partners pre-exist the relating, and the relating is never done once and for all."[16]

Combining these ideas, it can be claimed that humans in relation to dogs are hybrids as well. Thus, not only is the way in which the dog changes the human's behavior important, the reverse is also important as well. Latimer claims that humans are, and have always been, in "relationalities" with non-human species; she maintains that "human existence is never without life with non-human animals. This is not to stress the animal in the human, or other species as biological kindred, but to stress 'significant otherness,' in terms of interspecies socialities and interdependencies."[17] Franklin claims that humans and their companion animals form a "co(a)gency" where the animals' existence defines both the use of the space and the relations within. Thus, we understand the agency of the dog as a threefold concept, involving (1) dogs shaping the humans' behavior in co-being and co-doing, (2) both parties intentionally acting upon each other, and (3) dogs acting in relation to each other; or, as Latimer suggests, the question is "not just 'being-in-the-world' but a 'being-with' that includes other kinds."[18]

In our earlier analyses, we have distinguished two pedagogic approaches linked to a plausible understanding of agency for the dog.[19] They are constructed based on the role that the dog is given in the training process. In the dog-centered training approach, both positive reinforcement and negative reprimands are used as a means of conferring an active role upon the dog by asking what she desires the most, and by encouraging her to actively display the right behaviors in order to gain rewards. In the human-centered training approach, the basis lies in human authority, and hierarchical ranks structure the form of interaction.

The distinction between the two approaches can be reduced to the question of whether there are two individuals acting in cooperation, or a pack with a strong leader performing a given task. The agency of the dog is, consequently, built up differently in the two situations. With the dog-centered approach, the dog's individual abilities and desires are interpreted and taken into account. The dog as an active partner is given agency with contextual knowledge, and designated the ability to adjust or refuse to act in relation to certain consequences enacted on the part of the human. With the human-centered approach, the human defines the goals, and the forms of cooperation suggest that the dog lacks both agency and is restricted from engaging in its own moral codes of fair play. In both cases, however, the performance of the dog is a creation resulting from the trainer's abilities and actions, but the approaches differ in relation to the degree and nature of the assumed level of submission.

Latimer notes that co-constitutive relations are not just effective but also affective. They involve intermittent detachment and disconnection just as much as attachment and connection. However, according to Latimer, "This said, it should be emphasized that relations have no intrinsic value in themselves: the question is what the specificities of relations enact or displace."[20]

Data and Methodology

The analysis here is based on qualitative data consisting of 85 thematic essays written by dog trainers who actively engage in a variety of dog sports. Since our data collection questions were rather open in form, the informants were free to emphasize those subjects that were most important to them. The data were collected via dog-related Internet forums from August to September 2010. We asked informants to respond to eight thematic topics, covering their respective histories with dogs, training methods, everyday life and related problems, a description of the ideal dog and the ideal life for a dog, and the general acquisition of information about dogs and dog training.

The ages of the respondents varied from 20 to over 60 years. Most of them had had several dogs during their lifetimes, and more than half had at least two dogs at the time of data collection, with the biggest pack consisting of eight individual canines. The respondents represented an active, conscious, and educated group of dog handlers. They had acquired knowledge about training and canine behavior by reading books and articles, and participated regularly in training sessions and special courses.

There was a great variety in the breeds, ranging from Chihuahuas, Poodles, and Terriers to Afghan Hounds and Great Danes. The most common breeds were those that were most popular in dog sports: Border Collies, Smooth and Rough Collies, Australian Shepherds, Retrievers, Rottweilers, German Shepherds, Belgian Shepherds, and Doberman Pinschers. There was also a disproportionally high representation of less typical breeds, including the indigenous Finnish breeds such as the Lapponian Herder and Finnish Lapphund. The content of the data did not allow us to analyze the possible differences between caretakers with certain breeds in a statistical vein, but this was not our target in this study.

In this article, our earlier analytic work served as a foundation, even though the prior categories based on training methods were not accurate enough to fully consider the dog's agency. Therefore, for this analysis, we have re-read the data with our new theoretical objectives in mind and paid more attention to how the dogs' wills, intentions, and actions, as well as the dogs themselves, have been described both in training and in everyday life. We have analyzed the texts by using the method of dialogical thematization, reading the data in dialogue with the theoretical objectives. Our emphasis is on understanding the construction of the dog's agency in relation to the human's agency in the broader context of the construction of human–dog collective agency.

There are two main analytical themes: (1) the human-dog cooperation in training for the human-defined goals, and the dog's responses to the training, and (2) the animality of the dog and its consequences for understanding agency. These descriptions of these themes consist of the respondents' accounts of the dogs' qualities, the experienced success, challenges and difficulties faced in training, and the learning accomplishments and failures. The key sign is change, either in the trainer's actions or in the dog's. Looking back at Bekoff's understanding of dog-to-dog activity, such that one dog's chosen behavior has an effect on the other dog's actions,[21] the central question is about both morality and agency. In our analysis, we shall stretch this notion to human–dog interaction, asking how the dog's behavior changes the trainer's actions and vice versa.

Dog Sports as Context

The contextual space where we are examining the dog's agency is specific: humans and dogs entering dog sports and trials together. The specificity derives from the fact that dog sports, like any other sports, have fixed rules defining what to do and how to do it, and formal sports have national and international rules for official competitions and championships. Dog sports have become more and more popular in Finland during the past few decades; now, there are over 160,000 entries in dog sport trials on a yearly basis.[22] The most popular competitions include agility, obedience, tracking, and search and rescue. Dog sports have become so popular in Finland that both local canine clubs and commercial businesses have established heated training facilities around the country, from Helsinki to Lapland, in order to enable incessant training even during the cold, snowy, and dark winter months.

In every sport, the dog and the handler have to perform various predefined exercises, which are to be evaluated and scored by a judge. Most of the exercises include some form of obedience: heeling, retrieving, high jumping, going down, standing or sitting on recall, signaling exercises, and staying down for a demanded period of time. In agility exercises, the dog has to perform along a given track in a designated amount of time by following the verbal and physical signals of the handler; in search and rescue events, the dog has to follow a trail left by a human, or locate a hidden human based on traces of his scent. Many sports are a combination of different exercises testing the dog's motivation, endurance, and general physical shape. Most importantly, however, the cooperation between the handler and the dog is on display.

In order to advance from basic to higher levels and eventually to championships, the human must learn the rules and teach them to the dog. The rules include detailed descriptions of ideal performances, the principles determining how penalty points are judged, and the bases for dismissal. Considering the dog's agency, the most interesting principle is that penalty points do not only follow from incorrect actions during performances, but also from the quality of the cooperation between the dog and handler; the state of mind of the dog is also evaluated as well. It is stated in most of the rules that the dog has to show joyfulness, virility, and stamina during the performance. For example, here are some of the rules for obedience trials:

> The dog has to perform the tasks double-quick and controlled and express a clear wish to follow the orders and supervision of the handler. … When the dog reacts to the order or call immediately and displays clear willingness to perform the given task, it has to be given alpha unless there are mistakes or deficiencies in the performance.[23]

In the trial, the dog's eagerness to perform the tasks must be displayed by controlled bodily signs only: by wagging the tail, fastness, alertness, and sharpness during the performance. It is highlighted that signs of total submission to or fear of the handler will be punished in the most striking cases by dismissal:

> Reasons for dismissal can be, for example, help from outside, bringing toys or treats to the ring, punishing the dog, not following the given instructions, the running away of the dog from the ring, its aggressiveness, continuous and disturbing barking, uncontrolled behavior or clear unwillingness. Even the slightest punishing or hurting the dog will dismiss the whole test.[24]

From the point of view of agency, the situational conditions are strict for both the handler and especially the dog, since the latter has to learn such mental skills as self-control, tolerance of any stimuli, and other limits of the space while simultaneously displaying eagerness; this particular context thus offers an interesting space within which to analyze the dog's agency in interspecies interaction.

In Search of Agency: Humans and Dogs in Cooperation

Accomplishing the skills expected in any chosen sport is a lengthy process that typically takes years before the dog is "ready." This is why the presupposition among the trainers in our data was that the dogs they had purchased in order to enter dog sports should already have been willing to learn, eager to work together with the trainer, easy to motivate, and quick and trustworthy as partners in advancing in the sport(s) the trainer was interested in. The ideal situation was described as one where the trainer feels that the training is fun and joyful, with "perfect mutual understanding" and "one mind" being shared with the other in the world:[25]

> The best thing in dog training is reaching a perfect mutual understanding, and seeing how the dog enjoys cooperating with me.[26]
>
> The interest the dog pays in me is rewarding, as well as how quickly the dog learns, and [the fact] that all this is more fun than anything else.[27]

Therefore, the justification for goal-oriented training derives from the assumption of a shared reality and equal interests based on interpretations of the dog's gestures and behaviors. With this process of interpretation, the question about the limits of the dog's agency does not materialize, since the dog herself is defined as willing the actions. The companion dog is implicitly understood as having an inner need for training. Thus, training becomes defined as enabling the dog's own natural agency, providing a chance for species-specific activity and a better life:

> I am motivated by the dog's motivation. I take up practices with the dog, because it is so nice for the dog and for me.[28]
>
> Well, I think that also the Doberman Pincher yearns for training and borders; it might be frustrated if it would not be allowed to do stuff with its human.[29]

Following Latimer's thoughts, the question is now about "interspecies socialities and interdependencies,"[30] or as Sanders notes, humans and dogs cooperatively create an "interspecies culture" and assume a "couple identity" as they engage in collective action.[31] The agency, then, is constructed as co-dependent and collective.

However, this ideal is not always realized. A recurrent theme in our data was that the "perfect mutual understanding" and "joyful cooperation" had not been actualized to begin with, or had failed at being realized during some point of the training process. In this situation, the will of the dog contradicts that of the trainer; the dog simply communicates her unwillingness to cooperate. The dog may display her agency by refusal, thus forcing the trainer to change either her individual action or give up the entire goal:

> I can't practice any particular sports with the Great Dane because the dog does not enjoy any "unnecessary" tricks at all, and it can't be motivated by treats or toys, for example.[32]
>
> I have participated in agility camps also with the bitch, but she does not like to do the obstacles, so I have concentrated on training the male.[33]

The unwillingness of the dog to work with the trainer was interpreted by the trainers as the dog's insufficiency, comprised of passiveness, fearfulness, and refusal to cooperate. Thus, the dog still seems to be displaying her own agency, with her personal interests causing changes in the trainer's behavior. Applying the definition of animal agency by McFarland and Hediger, in that "choice is part of what defines agency,"[34] failing to train the dog due to her unwillingness is still an indicator of the dog's agency. Such behavior also causes critical self-reflection because it questions the human's ability to assume a "couple identity" with the dog. The dog may therefore bring the human to change herself:

> Sometime in the beginning, there were problems in the training. I totally blame my own inexperience. After a break, and through growing up and learning more about each other, we have been doing fine. We have found the common ground, and very well, too. ☺[35]

Thus, a failure in training was not fully attributed to the dog's qualities but to the problems in reciprocal interaction. The respondents highlighted the importance of learning to interpret the dog's state of mind: why she acts as she does, and what triggers her performance at each moment. Thus, the process is not just a question of finding the proper rewards, but one of finding out how the dog interprets the actions of the human, or how "my" behavior looks in the eyes of the dog. The dog as an agent has the capacity to change her actions and, even further, the self-image of the trainer.

The unwillingness of the dog to cooperate is not the only case where the dog's apparent "will" caused the trainer to change the future goals. There are examples in the data where the dog displayed her activity in a disruptive manner from the point of view of both the trainer and the set goals:

> Another problem, then, is the big ego of our dog, for when the first snarl is heard from the side, the dog will indeed answer back almost always and immediately. This problem has caused me to stop training. I do not dare to have my dog off leash.[36]

However, there were also examples of how the trainer did not give in to the dog's unwillingness, but kept demanding the correct action by inventing new means to motivate the dog, perhaps by taking courses and entering motivation clinics. Goal-oriented cooperation was ultimately seen as *the* way in which to be with the dog. When anything is done together, it is done to advance to the goal. In this context, the limits of the dog's agency are clearly defined by both the training needs and the human perspective:

> At home we use the so-called basic obedience, which calms the dogs down. When we do things in cooperation, we do it with full energy. When we don't, the dogs won't be noticed. I do not pet them when they ask for it. I call the dog to me, and pet it when it follows the call.

The ethos of doing things together shows that the training context was about mutually motivated action, but the human still defined the rules of cooperation for the dog to adjust or submit to. The boundaries of the dog's agency were therefore related to how human- or dog-centered the training's orientation was. However, in both cases, the course of action was justified by reference to what was best for the dog.[37]

The Animality of the Agency: "A Dog Is a Dog"

Training the dog for trials was predominantly portrayed as enabling species-specific activities for the dog. In addition, the data also contained notions stressing the need for the dog to just be allowed to "be a dog" on the basis that, in the end, the "dog is a dog." Thus, "being a dog" was recognized as being comprised of self-determination with respect to the focus and methods of actions – the dog was given a domain of agency of her own, allowed in environments thought as natural to the dog, including the countryside and forests outside of the human habitats. Defining the natural to the dog was based on its animality. Hence, the data showed constant implicit and explicit demarcations of how the dog's agency is best enabled, both in cooperation with the human and with minimal human control. We interpreted this tension in regard to whether and to what extent the idea(l) of "perfect mutual understanding" was the motivating and justifying goal when the dog was also seen as "a dog" whose animalistic needs and possibilities must be equally accounted for. In these interpretations, the short Finnish history of urbanization can be seen as forming a question of how far the dog's "natural" agency is positioned as part of human-defined social contexts:

> I think that the dog is happiest in the countryside or countryside-like environment, where it has the possibility to do things without constant ordering and supervision, and, of course, to romp around as much as it ever wants.[38]

The "being a dog" also contained possibilities to interact with other dogs and to form social relations with them. It was frequently emphasized that the dog should not be excessively anthropomorphized, nor should it be made responsible for human issues that are not in the dog's natural domain. While the dog was seen as a part of the human world, and the sharing of a mutual reality was emphasized, the agency of the dog was nevertheless seen through the context of her animality. The positioning of the dog on the fringes of both society and nature was especially visible in the context where the limits of agency were defined:

> I believe that part of the good life for the dog is also a possibility to have close relations with other dogs. A happy dog has some task which it performs together with the human. Be it guarding or training dog sports … but a happy dog may also be a dog: it has not been humanized too much, and it has not been given responsibility over things that it can't be responsible for, like raising human children, receiving guests, and leading walks.[39]

The spontaneous action of the dog may well be considered problematic in the training context. However, when the dog was expected to differentiate between degrees of threat to the humans and act accordingly, spontaneity was the required trait. Even though the respondents strictly wanted to avoid anthropomorphizing, the dog was also expected to accurately interpret the human social sphere. In these situations, the "natural" agency of the dog that was interpreted as benefiting the human was recognized and even desired.

The animality of and in the dog – and the responsibility of the human to provide for it – was constantly emphasized. However, in everyday situations, hostility and aggression towards other dogs and humans or fear of unfamiliar things were not only considered as strongly undesired but also as influenced by human agency. Walking a dog that barks and growls at other dogs creates pressure and expresses the failure of shaping the dog's behavior, demanding constant efforts on behalf of the human:

> My elder dog's problem was on-leash aggression and she [still] does not particularly like other bitches and is aggressive towards them. Growling, barking, and lunging were controlled by punishing the dog for wrong behavior and rewarding the dog for the right actions, even though it took quite a long time (four years).[40]

In relation to society, humans have the responsibility to control the dog's animality. In this context, the human has both the right and the duty to define the limits of the dog's agency. Based on the criteria for good behavior, the dog was placed in the human world with requirements thus related to the dog's social agency instead of her animality.

Conclusions: The Tensions in Agency

The agency of the dog is constructed situationally. The dog has the ability to change the trainer's behavior, emotional state, and actions from joyful co-acting to disappointment and giving in. The trainers and their dogs constitute a co-agency of two distinct individuals in an "authentic social exchange,"[41] where both participate in negotiating the course and quality of actions, "consciously behaving so as to achieve defined goals," as Sanders notes.[42] The construction of co-agency becomes highlighted in dog sports and competitions, where, by definition, the human and the dog are expected to work together as one.

Considering agency from the point of view of being a political subject includes the possibility to self-determine both the objects and procedures of action.[43] At the core of being a human citizen are the initial possibilities to participate in the issues important to the subject, and to express views in a way that seem feasible to him/her.[44] Political agency highlights the self-determination of an individual concerning his/her actions, not the effectiveness of agency as such. Regarding dogs, the question of being a political agent culminates in the tension between animal nature and societal nature, leading to the question: What kind of actions and agencies must be controlled, and when and how can they be?

In dog sports, it is not only the moral codes that define the agency of the dog; it is the individual human deciding how the dog must act and what she must want. Highlighting cooperation shows that the will of the trainer must become that of the dog, too. The happier the dog is interpreted to be during a performance, the more the action is understood as a sign of the task "internally" desirable to her. Then, training the dog becomes justified by the definition of the dog's own will and well-being; in other words, a good trainer wants to offer the dog a more meaningful life. In this process, the representations of the "best" for the dog and "best" for the human overlap seamlessly. However, it is difficult to know where the dog's interest actually lies: in cooperation itself, the reward after, or both. Or maybe even somewhere else.

Dogs are pack animals with a need to understand the surrounding world and recognize causalities, a need that is necessary to all biological beings.[45] Thus, goal-oriented training at its best brings this rationality to the dog's world. She learns how she, through her own actions, can have an effect on what happens to her, and perhaps gain acceptance, treats, and play. It does not matter if the human simultaneously accomplishes recognition in competitions, which in the dog's world most likely does not factor into existence.

Goal-oriented training with its specific causalities does not necessarily mean that the dog's agency is repressed. As even human agency varies, reflecting the conditions of time, space, and power relations as we noted in the beginning, the agency of the dog seems to concur with this conception – perhaps not straightforwardly, but as an interspecies interpretation of those conditions. The dog has the capability to express her agency by consenting willingly to the interaction, or by refusing to do so. In both cases, it is co-constituting reciprocal interspecies agency.

Notes

1 Staats, Wallace, and Anderson, "Reasons for Companion Animal Guardianship."
2 Charles, "'Animals Just Love You as You Are,'" 3–4.
3 Koski, "Koira jälkimodernissa yhteiskunnassa"; Koski and Bäcklund, "On the Fringe."
4 Möller Gabrielsen, "'Farlig hund.'"
5 Koski and Bäcklund, "Koiran virka"; Koski and Bäcklund, "On the Fringe."
6 Power, "Furry Families," 552.
7 Koski and Bäcklund, "On the Fringe."
8 Bekoff, *The Emotional Lives of Animals*, 89.
9 McFarland and Hediger, "Approaching the Agency of Other Animals."
10 Ibid., 18.
11 Koski and Bäcklund, "On the Fringe."
12 For example, McNay, *Gender and Agency*.
13 Åkerblad, "Epävarmuuden tuolla puolen," 45.
14 Sanders, "Actions Speak Louder Than Words," 214.
15 Latour, *We Have Never Been Modern*.
16 Haraway, *The Companion Species Manifesto*, 12.
17 Latimer, "Being Alongside," 78.
18 Franklin, "'Be(a)ware of the Dog'," 90.
19 Koski and Bäcklund, "On the Fringe," 24–44.
20 Latimer, "Being Alongside," 80.
21 Bekoff, *The Emotional Lives of Animals*, 89.
22 Finnish Kennel Club, "Koiraharrastukset."
23 Suomen Kennelliitto, "Tottelevaisuuskokeen säännöt ja koeohjeet," 25.
24 Ibid.
25 Latimer, "Being Alongside," 80.
26 E 29. Refers to the serial number of the essay in the data.
27 E 35.
28 E 62.
29 E 72.
30 Latimer, "Being Alongside."
31 Sanders, "Actions Speak Louder Than Words," 420.
32 E 65.
33 E 79.
34 McFarland and Hediger, "Approaching the Agency of Other Animals," 18.
35 E 75.
36 E 73.
37 Koski and Bäcklund, "On the Fringe."
38 E 12.
39 E 51.
40 E 62.
41 Sanders, "Understanding Dogs," 211.
42 Ibid., 207.
43 Arendt, *The Human Condition*.
44 Rosanvallon, *Counter-Democracy*.
45 Bradshaw, *In Defence of Dogs*.

Bibliography

Research Materials

All thematic essays are in the possession of the authors.

Literature

Åkerblad, Leena. "Epävarmuuden tuolla puolen. Muuttuvat työmarkkinat ja prekaari toimijuus." PhD diss., University of Eastern Finland, 2014.
Arendt, Hannah. *The Human Condition*. Chicago: Chicago University Press, 1958.
Bekoff, Mark. *The Emotional Lives of Animals*. Novato, CA: New World Library, 2007.
Bradshaw, John. *In Defence of Dogs*. London: Allen Lane, 2011.
Charles, Nickie. "'Animals Just Love You as You Are': Experiencing Kinship across the Species Barrier." *Sociology*, 48 (2014), 1–16. doi: 10.1177/0038038513515353.
Finnish Kennel Club. "Koiraharrastukset." Accessed October 2, 2014. www.kennelliitto.fi/koiraharrastukset.
Franklin, Adrian. "'Be(a)ware of the Dog': A Post-Humanist Approach to Housing." *Housing, Theory and Society*, 23 (2006): 137–56.
Haraway, Donna J. *The Companion Species Manifesto*. Chicago: Prickly Paradigm Press, 2001.
Haraway, Donna J. *When Species Meet*. Minneapolis, MN and London: University of Minnesota Press, 2008.
Koski, Leena. "Koira jälkimodernissa yhteiskunnassa." In *Merkityksen pohjasäikeet*, edited by Päivi Kosonen and Tuomas Nevanlinna, 49–70. Helsinki: Tutkijaliitto, 2006.
Koski, Leena, and Pia Bäcklund. "Koiran virka: susi kaupungin kaduilla vai ystävä erämaassa?" *Alue ja ympäristö*, 41 (2012): 22–33.
Koski, Leena, and Pia Bäcklund. "On the Fringe. The Positions of Dogs in Finnish Dog Training Culture." *Society & Animals*, 23 (2015): 24–44.
Latimer, Joanna. "Being Alongside: Rethinking Relations amongst Different Kinds." *Theory, Culture & Society*, 30 (2013): 77–104.
Latour, Bruno. *We Have Never Been Modern*. Cambridge, MA: Harvard University Press, 1993.
McFarland, Sarah E., and Ryan Hediger. "Approaching the Agency of Other Animals: An Introduction." In *Animals and Agency: An Interdisciplinary Exploration*, edited by Sarah E. McFarland and Ryan Hediger, 1–20. Leiden and Boston: Brill Academic Publishers, 2009.
McNay, Lois. *Gender and Agency: Reconfiguring the Subject in Feminist and Social Theory*. Cambridge, UK: Polity Press, 2000.
Möller Gabrielsen, Ann. "'Farlig hund' som diskursiv konstruktion. Disciplinering av hundar och hundägare i Norge." *Elore*, 20 (2013): 83–105.
Power, Emma. "Furry Families: Making a Human-Dog Family through Home." *Social and Cultural Geography*, 9 (2008): 535–55.
Rosanvallon, Pierre. *Counter-Democracy: Politics in an Age of Distrust*. Translated by A. Goldhammer. Cambridge, UK: Cambridge University Press, 2008.
Sanders, Clinton R. "Understanding Dogs: Caretakers' Attributions of Mindedness in Canine-Human Relationships." *Journal of Contemporary Ethnography*, 22 (2003): 205–26.
Sanders, Clinton R. "Actions Speak Louder Than Words: Close Relationships between Humans and Non-Human Animals." *Symbolic Interaction*, 26 (2004): 405–26.
Staats, Sara, Heidi Wallace, and Tara Anderson. "Reasons for Companion Animal Guardianship (Pet Ownership) from Two Populations." *Society & Animals*, 16 (2008): 279–91.
Suomen Kennelliitto. "Tottelevaisuuskokeen säännöt ja koeohjeet." Accessed February 25, 2015. www.kennelliitto.fi/sites/default/files/media/toko_saannot.pdf.

2 Human and Non-Human Agency in Icelandic Film

Of Horses and Men

Stella Hockenhull

Introduction

Partway through Benedikt Erlingsson's directorial debut entitled *Of Horses and Men* (*Hross I Oss* 2013), one of the central characters, Kolbeinn (Ingvar E. Sigurdsson), rides his horse through the remote Icelandic landscape. Appearing proud and erect, he negotiates the wild and inaccessible countryside; showing off the mare's paces, he and his horse seem in harmony with each other and with nature. This sequence is one of a number of mini-narratives that comprise the film; each commences with an extreme close-up of an Icelandic horse's eye within which a human reflection is mirrored, thereby demonstrating a convergence of spirit between man and beast. Indeed, *Of Horses and Men* is filmed predominantly outdoors, and centers on the relational agency between human, animal and landscape.

However, the film differs from most which focus on animals in that ordinarily they are treated anthropomorphically, in other words ascribed human characteristics and emotions enabled through the style and language of the film. Anthropomorphic treatments, dependent on digital means, editing and cinematography, offer a particular interpretation whereby the animal appears to perform and operate as "conscious intending subject," a state termed "disnification" by Steve Baker, which he suggests follows

> a kind of pattern: when the animal is put into visual form, it seems somehow to incline towards the stereotypical and the stupid, to float free from the requirements of consistency or of the greater rigour that might apply to other non-visual contexts. The image of the animal seems to operate here as a kind of visual shorthand, but a shorthand gone wrong, a shorthand whose meanings intermittently veer from or turn treacherously back upon that of the fuller form of the text.[1]

This type of representation ensures that the narrative pull leans towards sentimentality, and, for Baker, renders the animal human characteristics.

Erlingsson completely reverses this practice, denying the horses in his film any personification, yet nonetheless spotlighting them, and providing them with greater moral agency and judgment than their human counterparts. Furthermore, the director uses the geography and climate of this harsh northern landscape as its own agent and driving force in order to advance the narrative. Indeed, indirectly the director bestows the setting of the film with its own agency, a dynamic that ecological geographer Val Plumwood argues need not be confined to human or human-like consciousness.[2] This essay analyzes the interrelationship between the Icelandic horse, human behavior and the landscape in *Of Horses and Men* within the context of the country's geography. It argues that, by avoiding an anthropomorphic treatment, the director offers greater currency to the non-human animal, acknowledging its right to idiosyncrasy and choice within its natural environment.

Animal Agency and Representation in Film and the Media

Before analyzing equine representation in Erlingsson's film, however, it is important to discuss the ways in which animals have traditionally been discussed in film and the media. Character animals have generally been integrated in the study of human characterization and, in Classical Hollywood cinema particularly, there is typically an individualized and imaginary psychologically rounded character who is developed in some depth, and who possesses identifiable aims and objectives, thus enabling the narrative to progress. David Bordwell and Kristin Thompson have discussed this in detail, suggesting that "historically [--] the cinema has tended to be dominated by a single mode of narrative form [--] we shall refer to this dominant mode as 'the classical Hollywood cinema.'"[3] This approach specifically depends on the supposition that characters are causal agents, whereby the narrative centers on personal psychological characteristics, choices and individual traits. These characters are usually human and, in turn, stars,[4] but can also take the guise of animals or robots, and even household objects and landscapes.[5]

Nevertheless, as noted above, unlike humans, and usually in order to advance the plot, animal characters are anthropomorphized and personified which enables spectator empathy and identification. In a similar vein to humans, to mobilize narrative progression, character animals experience difficulties, and usually the sequence of events commences with equilibrium and stasis, and a series of problems create disequilibrium before stability and closure ensues. To muster sentiment and to understand character motivation, cinematic techniques such as the close-up and point-of-view shot are used to ask the spectator to adopt the animal's standpoint.

Unlike Bordwell and Thompson who do not explore the animal protagonist as motivating character, Jonathan Burt considers its representation in film. He suggests that when the animal look merges with the camera lens "film effectively turns the animal into a camera, a non-human recording mechanism."[6] Burt is discussing Robert Redford's 1998 film, *The Horse Whisperer*, but he does not attribute the horse with human qualities; he merely asks the spectator to identify with the animal's trauma through point-of-view shots which further the narrative, thus suggesting that the animal is afforded agency.

Sometimes, the filmic animal speaks human language and is provided a personality, specifically in Disney cartoons for example; this anthropomorphism imparts the most common form of animal integrity, and may promote it to star. When this occurs, the onscreen animal is also heroized through their performance capabilities, often staging daring feats to complete a rescue operation and redeem a situation. This notion of animal stardom has been extensively researched by the media sociologists Emmanuel Gouabault, Annik Dubied and Claudine Burton-Jeangros. Termed "superindividuals," such animal figures, Gouabault, Dubied and Burton-Jeangros argue, have become increasingly personified in Euro-American information media and, moreover, are frequently described as human characters and given human names, with their feelings and thoughts described in some detail. Even in wildlife films, they suggest, the animal is usually anthropomorphized, and wild and endangered species can become flagships of all threatened animals. It is this treatment which permits such starification, yet also, for them, introduces unsatisfactory connotations denying the animal its animality.[7]

Whereas the above theoretical paradigms focus on personification, they do not discuss the notion of animal performance in film. This is a thorny issue, complicated because film language often eliminates the possibility of this type of analysis. Nonetheless, in her seminal article on screen performance, Brenda Austin-Smith examines three films, one of which includes the performance of the donkey named Balthazar in Robert Bresson's 1966 film, *Au Hasard Balthazar*. Adopting Michael Kirby's system for the analysis of different types of acting, she proposes that animals in film are capable of "simple acting," and that, "in order to be valued, performance must be noticed and identified *as* performance, rather than as star exhibition, an artefact of editing or the traces of someone merely living in front of the camera."[8] In her analysis she draws upon Kirby's continuum of acting and non-acting, which he first outlined in 1972. In what he terms non-matrixed performance, there is no intention to portray a character, but, at the other end, performance is multidimensional and the actor portrays a number of specific emotions to create the pretense. Somewhere between non-acting and complex acting simple acting occurs, whereby the performer does something to replicate or impersonate a character. On these occasions there is an intention to act on the part of the performer, although no emotion needs to be involved.[9]

For Austin-Smith, on the surface, the donkey's performance does not constitute simple acting – "the donkey and the character are minimally distinguished."[10] Furthermore, the creature does not and cannot feign, simulate or impersonate. However, on Kirby's matrix, the animal's freedom to make choices, the right to be idiosyncratic, awards it the status of simple actor, and this is made obvious to the onlooker through the close (and sometimes knowledgeable) observation of its external bodily signals. If, as Austin-Smith argues, the consequences of actors' choices are "visible performance signs,"[11] then this concept can be applied to animals, a factor that also aids in the construction of key traits which create characters. In sum, according to Austin-Smith, "movies give us fictional beings we figure out by watching them respond to a world arranged by someone else."[12] Nevertheless, even if it is accepted that animals cannot feign or impersonate, then the choices that they make elicit distinctiveness, and this is inadvertently presented outwardly, even though it is often necessarily constrained by the film language.

Despite the above, very little has been written on the notion of animal agency in film, albeit there is a wealth of material available on live animal performance. Theatre scholar and practitioner Michael Peterson has written on the subject and argues that an anthropomorphic reading of animal performance renders it void and "reduced to a sign, [whereby it] contributes nothing to performance but expense and inconvenience."[13] David Williams' less conventional and more challenging theory specific to the analysis of horse performance permits the animal its own integrity by proposing an affirmative symbiotic relationship between human and non-human animal. He argues that only a complete comprehension of the mutually beneficial affiliation between the two enables a true judgment of the horse's intervention. Williams suggests that there exists a rapport between human and animal based on the concepts of what he terms "answerability" and "responsibility," thus acknowledging animal "thinking" with body.[14] Performance theory thus offers the notion of animal agency minus, to a certain extent, anthropomorphism, whereas a film and media analysis of animals tends towards their personification.

Human and Non-Human Performance in *Of Horses and Men*

The Iceland horse, which straddles the human world and the wilderness, the mythical and the real, is dignified, never disnified in Erlingsson's film, nor does the director mobilize the notion that these animals possess characteristics similar to humans. *Of Horses and Men* represents the animals as sentient beings, but at no point does the filmmaker make the horses "speak" for themselves, either visually through the editing of the film, or by imposing them with human voices. Indeed, the horses in this film do not have an active look that motivates the plot, and the director does not provide point-of-view shots or shot reverse shots to present their perspective. Their eyes, in fact, frequently conceal any means of interpretation because of the human reflections retained within, yet they function in what Austin-Smith terms a "symbolized matrix" and are purposeful and idiosyncratically goal-directed, operating independently and with freedom of choice.

This is exemplified in the above-mentioned opening sequence. Prior to Kolbeinn's ostentatious equine display, the film commences with an amplified image of a horse's eye, the figure of Kolbeinn reflected within. A further edit in medium shot reveals that the eye belongs to a horse, a grey mare named Grána. She is Kolbeinn's pride animal and he is getting her ready to display her paces to his fellow horse breeders, an occupation which is a way of life in Iceland. Subsequently, Kolbeinn, a well-dressed man, approaches her and speaks quietly and lovingly, telling her to "be a good girl now – be still my darling." Grána is clearly an object of his affection, albeit initially she counters his advances, refusing to be caught and moving away from him in an attempt to escape.

Nonetheless, Kolbeinn perseveres and manages to corner her, and, when he eventually succeeds in her capture, he talks gently and caresses the mare's face and neck, his whole demeanor and behavior appearing as that of a persuasive lover. At this point, however, she spurns his advances and, operating as a free spirit, rears upwards and out of his reach, temporarily escaping. Kolbeinn is finally successful in placing a bridle over her head, although clearly the mare has never been broken and, to undertake this process, he proceeds slowly, stroking her neck and back. Although it appears as though Kolbeinn has instigated the change in Grána's behavior, the mare introduces her own actions to be interpreted by the spectator within the context of the narrative. Indeed, Grána might be perceived as acting within a "symbolized matrix," purposeful and goal-directed, operating independently and with free choice. The mare produces a performance for the spectator, not as an actor, or one that can be considered anthropomorphically, but as a horse whose options can be interpreted through an understanding of animal behavior, notwithstanding that she is constrained to an extent by the language of the film.

The next shot introduces the human intrigues and relationships which form part of the narrative in *Of Horses and Men*: Solveig (Charlotte Bøving) is a widow and, amongst a number of other women who live on the island, an admirer of Kolbeinn. She and her family reside on a small farm, and clearly have been informed that Kolbeinn is to display the paces of his newly broken mare on this specific day, even though such narrative information is withheld from the spectator. She along with her mother and son all watch intently from the window of their house as horse and man trot proudly past along the road. Kolbeinn glances conceitedly across at the gathering and smiles because it is clear that they all admire his horse. Here, Erlingsson introduces the theme of voyeurism, which has, as noted, been initiated through the close-up of the animal eye. Solveig's mother watches Kolbeinn through a pair of binoculars and a succession of shots produces a series of onlookers from different farmsteads, all of whom are horse breeders or work with horses. Through point-of-view shots, the spectators attentively watch Kolbeinn and Grána's display, presumably keen to assess the horse's prowess. This suggests not only a perception of the country's insularity whereby Erlingsson presents the Icelandic men as inquisitive and jealous and the women as predatory and desperate for marriage, but also signifies the importance of the horse to the Icelandic community.

Seated erect, his head held high, Kolbeinn takes pleasure in her strides, particularly the unique gait known as the tölt, a pace which Icelandic horses are renowned for.[15] At one point the pair is visible against the remote backdrop of the Icelandic landscape where gently rolling hills and blue cloudy skies form the scenic background. Kolbeinn stops and the camera frames him in medium shot, the spectator now privileged a spectacular image of snow-capped mountains with small houses nestled at their foot. Here, the cinematographer uses a low angle shot to ennoble man and beast and to emphasize the notion of horse and rider in harmony with the landscape. Indeed, from the beginning of the film, and throughout the story, the Icelandic horse is visually aligned with its surroundings, in which it toils to survive. At this moment Kolbeinn sings with happiness, an uplifting response as he travels along to the sound of the horse's hooves on the metal of the road. Solveig and her mother and son all stand on the doorstep to watch as the pair arrives.

Kolbeinn's pride in his horse is apparent and the mare cooperates accordingly. When asked, she increases her speed and the spectator can only guess that she is destined to be his prize winner, Kolbeinn's competitive nature indicated by the array of medals and prizes for his riding expertise and horsemanship on display in his house. Throughout the sequence, the camera focuses on the animal's legs leaving the audience in no doubt that it is this beast of burden which is of prime importance to the island. Indeed, Grána's pace is smooth and horizontal, a style which is a source of pride for the Icelanders; even Solveig's mother is heard to comment on the animal's "beautiful gait." On arrival at Solveig's house, Kolbeinn loosens the mare's girth and is invited in for coffee. He subsequently enters and, significantly, Erlingsson returns centrality to the animal, and the human voices now become distant indicating that the narrative propulsion has been transferred to the horses. At this juncture the focus remains on Grána who has observed the couple enter the house, before the camera transfers to follow the interaction that ensues between the mare and Solvieg's horse, a black stallion named Brúnn.

Brúnn is clearly attentive to Grána. He walks the length of the fence and snorts and whinnies, a partial display of sexual attention in horses. The mare, distracted yet alert to his interest, becomes agitated and attempts to break free of her tether. Meanwhile, Kolbeinn and Solveig are seated inside and diegetically seem oblivious to the sexual attraction between the two animals. Erlingsson's placement of the characters within the confines of the house but viewed from outside, and the use of the indistinct sound of their voices, suggests that they are no longer the key narrative agents or capable of mobilizing change, a factor evidenced later in the sequence.

Presently, Kolbeinn departs Solveig's household with a friendly wave, unaware that Grána and Brúnn's instincts have proven too strong. What follows is a significant and shocking experience for Kolbeinn, his onlookers and the spectator. Despite Kolbeinn's protestations, the stallion mounts the mare with the man on board and proceeds to mate. The act is carefully choreographed, and involves actual pairing. Described by Calum Marsh in his critical review of the film as an "impromptu sex act [which] is a stand-in for what Solveig wishes Kolbeinn would do to her,"[16] the spectator is never privy to human or, obviously, animal emotions in the film. Instead, the animals make their choices, not as actors, but through what Austin-Smith describes as

> "an air of knowing unknowness" [which] hopelessly and wonderfully complicates the relations between choice, self-knowledge and performance. … The performer's actions and responses are memorable because we receive them as the products of character decision and response rather than as gestures intended by an actor to produce mere consistency.[17]

The animals here present an "air of unknowness" in their action, because they cannot operate as actors, feign or pretend, and neither are their acts received as products of character decision. Notwithstanding that their choices are shaped by the filmmaker, they perform with idiosyncrasy. In this respect they show us all that they are, yet their actions present narrative intervention enforcing Kolbeinn to make decisions he would not otherwise have made.

The mating scene between Grána and Brúnn is witnessed and framed through the binoculars of the shocked onlookers including Solveig, who claps her hand to her mouth in disbelief. Kolbeinn loses his hat which he had donned so carefully at the beginning of his ride, and narratively he has lost his dignity and, vicariously, his companion. Mating complete, he peers at the surrounding houses aware that he must surely have been seen; given the man's attitude and pride in the mare, the suggestion is that he has been cuckolded, and in full view of the islanders.

In *Of Horses and Men*, Erlingsson's horses are never fully bound by human direction, neither are they submissive beings on which humans mark meaning. Instead, the director provides sanction to the horses which contribute to the order of narrative space through their own creaturely behavior. Indeed, as Chris Philo and Chris Wilbert argue,

> In our view, it is also vital to give credence to the practices that are folded into the making of representations, and … to ask how animals themselves may figure in these practices. This question duly raises broader issues about non-human agency, about the agency of animals, and the extent to which we can say that animals destabilize, transgress or even resist our human orderings, including spatial ones.[18]

Thus, the Icelandic horses in *Of Horses and Men* effect change, both through their own agency and in the narrative. That they "resist human orderings" is not necessarily true, as Erlingsson wrote this scene into the script. However, their innate and idiosyncratic actions also inform narrative meaning. Further, their natural environment offers them freedom, a tradition of the past which still exists.[19] As Kirsten Hastrup remarks, they "belonged to a separate space; they were tamed yet left far more or less wild and out-of-doors for the entire year. They had both will and character, and I have here designated them as metaphorical humans."[20] While Erlingsson "controls" their activities through the film language, to a certain extent the natural behavior of the animals also invariably destabilizes his work.

From this ill-fated episode, the punishment of Grána commences. Filmed from a variety of angles, Kolbeinn is witnessed riding through the landscape, now, presumably through mutual embarrassment and discomfiture, conspicuously devoid of the inquisitive onlookers. On his return home the camera cuts to an interior shot as he obtains a gun. Seen from inside the house, Kolbeinn raises it at the mare, takes aim and shoots her. This act throws him into grief and despair and, after firing a second shot, he slouches forward, bent double in apparent sorrow.

Subsequently though, Kolbeinn expresses little sentiment, and there is sparing dialogue throughout the scene, even though this also serves to reinforce the equality of human and animal in the film, if not the predominance of the horse in the narrative. Indeed, the director's avoidance of sentiment succeeds in collapsing human–animal boundaries to suggest visceral passion in both species.

Human Folly and Equine Vision

Subsequent to this episode, a second mini-narrative commences with an extreme close-up of a horse's eye. As Burt argues,

> the image of the animal's eye reflects the *possibility* of animal understanding by emphasizing animal sight. This does not mean that the eye gives any access to what is understood but it does signal the significant participation of the animal in the visual field.[21]

Although the spectator cannot know what the animal thinks or feels, Erlingsson creates space for equine participation in this "visual field." He fashions a parallel whereby the human and animal see the world from different perspectives, suggesting a possible analysis of the horse's mind. On this occasion the horse is named Jarpur, and again an image of his owner, Vernhardur (Steinn Ármann Magnússon), is reflected within the animal's eye at the outset. Unlike Kolbeinn, Vernhardur is an unkempt and foolish man, and a desperate alcoholic. In medium shot, he is first seen driving his vehicle across the rough terrain, an image which is shown alongside a horse's head, the latter viewed in profile grazing. Vernhardur's aim is to swim to an offshore Russian boat astride his horse to procure cheap alcohol.[22] He mounts the animal and the pair advances towards the boat in freezing conditions, the horse swimming with his head barely visible above the water, his master astride. The cold water makes the feat almost impossible, and the Russian sailors appear disbelieving of the man's folly.

However, man and horse arrive safely on board but, ignoring warnings from the Russian crew about the strength of the liquor, on his return to dry land Vernhardur drinks copious amounts. Throughout this performance the man is casually observed by his horse, although the vocabulary of Erlingsson's film invokes no anthropomorphic comment on the man's folly. Indeed, the director ignores traditional techniques used in filmmaking to enable spectator understanding of the characters' thoughts; by using little or no dialogue and few close-up or point-of-view shots of either human or animal, at no time is the spectator enabled Vernhardur's inner emotion or the "thoughts" of the animal.

Watched over by Jarpur, Vernhardur presumably develops alcohol poisoning and subsequently dies and, at this moment, the animal drops his head to investigate. As Austin-Smith would argue, this behavior is part of the symbolized matrix whereby the filmmaker makes choices concerning the film language, yet the spectator is not coerced into an anthropomorphic reading. Instead, Jarpur is unable to behave in any way other than the animal that he is and Erlingsson does not rob him of this capacity. The sequence ends with a medium shot of the horse standing over his dead master, the ship in the background and the wild landscape providing a milieu to human folly and non-human existence.

The next sequence returns to Kolbeinn, who is tending to his dead "unfaithful" horse, Grána. He strokes her head and arranges her tail almost as a sacrificial offering to the gods. Seen from an overhead perspective, the mare is laid out still wearing her saddle and bridle and, in a subsequent lengthy shot, he tenderly strokes her and, placing his head close to hers, states "be good." Grána's death is juxtaposed with that of Vernhardur's demise, Erlingsson using an edit to the funeral to link the two events. Indeed, throughout the film, death is a significant theme occurring naturally and unnaturally because of the harsh weather conditions and inaccessible and hostile landscape.

Non-Human Agency and the Icelandic Landscape

As noted, if the horse becomes a significant instigator of change in *Of Horses and Men*, then the landscape is also provided agency. Plumwood argues that, in terms of the cultural landscape, humans are invariably seen as focal agents and this interpretation overrides all forms of non-human agency. As she explains, "To describe the land as a 'landscape' is to privilege the visual over other, more rounded and embodied ways of knowing the land, for example, by walking over it, or by smelling or tasting its life, from the perspective of predator or prey."[23] While Plumwood is chiefly arguing on ecological grounds, she proposes that "we do need to question systems of thought that confine agency to a human or human-like consciousness and refuse to acknowledge the creativity of earth others, whether organized into a single system or not."[24]

The northernness of Iceland is important in motivating the narrative in *Of Horses and Men*, operating not merely as setting but also as non-human intervention. Iceland is an inaccessible country surrounded by sea, with only small tracts of land habitable, which lie mainly in the southwest. Accordingly, its main population is based in the most northerly capital in the world, Reykjavik. This results in the remainder of its society isolated, with only a scattering of towns and rural farmsteads. Historically, these outlying regions often formed the center of their own self-contained worlds, with the occupants keeping animals both as transport and as a means of survival.

Furthermore, the Icelandic horse has existed there for hundreds of years and is now a prominent symbol and feature of this landscape. Initially introduced by the Vikings, the horse played a major part in society from the ninth century onwards, and was kept for meat, riding and transport. The animal's breeding and resilience meant that it was able to carry out a variety of arduous tasks without difficulty and, as Hastrup suggests, "Horses were an object of much pride. They were perceived as strong, so strong that they could carry two barrels of iron, beer, or butter at the same time."[25]

It is this geography and the country's inhabitants that Erlingsson uses to effect change and interject meanings into the narrative. Frequent shots of the harsh environment are interlaced with images of the people and horses, and these are often depicted in, what Martin Lefebvre terms, "space freed from eventhood." Lefebvre distinguishes between setting and landscape in cinema, and, for him, customarily setting forms part of narrative space which includes spaces that are part of larger unseen locations. This is frequently presented as disjointed because it is divided up into action, shot and frame, whereas landscape, on the other hand, is "*space freed from eventhood* (for example war, expeditions, legends)."[26] In this mode, it is placed for spectator contemplation and need not form part of the film's diegesis. When landscape is used to motivate the narrative it is permitted its own agency and performs an aesthetic as well as a functional role. In this film, the Icelandic landscape becomes an intervention in events operating as a cause of hardship and death. Indeed, a series of deaths and funerals punctuate the film.

The final sequence of events in the film culminates in an annual task undertaken by the Icelanders for the winter months: the corral of the wild horses takes place in the autumn when farmers and helpers gather together in order to protect the animals from the harsh weather and to assess the herd. Again the event commences with an extreme close-up of Brúnn's eye with Solveig mirrored within. The animal has now been castrated as part of a silent agreement between Kolbeinn and Solveig in recognition of the animal's transgressive behavior. Solveig dons lipstick for the occasion and bids farewell to her mother and son before setting off on the adventure, and to join the group of riders who have volunteered to help with the task. They all move off to commence the horse roundup, sociably passing a hipflask of alcohol between them. Emphasizing the sturdiness of the horses, the sequence commences with a series of underwater shots of the animals' legs as they cross rivers and rough terrain, all swimming strongly and progressing at a jaunty pace. A little later, a decision is made to divide the group and to send the best two riders via the more difficult route. Mention is made that this is a tradition that has existed for over 1000 years, possibly as both a ritualistic and fertility exercise. Kolbeinn and Solveig are selected to take the more arduous way and, after an exchange of glances intimating complicity, they set off across a remote section of the island. Judging themselves hidden from sight they partially remove their clothing and copulate. Brúnn and Kolbeinn's new horse graze quietly by, the irony of this situation not lost on the spectator.

The film ends when Kolbeinn and Solveig return to their duties in the roundup, and at this point all the characters converge, and Kolbeinn selects another horse, a grey mare, presumably to replace Grána, and he and Solveig move towards each other indicating that there are prospects for their romance too. The camera tilts upwards, and the spectator is enabled an overhead shot of the archway through which the horses are driven. The film's ending concludes here, suggesting the lifecycle of the horses, the annual rotation for the Icelanders and, in turn, the durability and unchanging nature of the land. Death is part of the Icelandic reality, and the people and their horses coexist in the sometimes difficult conditions.

Erlingsson creates an observation of this yearly round, and while the story has a loose narrative and is told from a human perspective, the horses form the structure and focus of the film.

Conclusion

Unlike most films that either humanize or ignore the animal, in *Of Horses and Men*, Erlingsson makes no attempt to assign them human traits and thoughts through the film language. Never does the camera operate from their point of view through editing, although, at times, it does create space for the animal to "participate in the visual field." Conversely, rather than becoming what Carol Adams terms "absent referents,"[27] the equine representation enables subversion, violation or even a dispute of our human orderings. Furthermore, the conduct of the horses exposes human struggle and folly in this isolated and untamed world. *Of Horses and Men*, as the prioritization of the animal in the title indicates, although this actually translates as "Of Horses in Men," awards the horse primacy, and its identity and appeal are partly wrought from its fusion with its mythical yet sometimes hostile environs. Ultimately, the film captures the horses as the wild, graceful and independent beasts that they are, an aspect which integrates animal, landscape and community.

Notes

1 Baker, *Picturing the Beast*, 175.
2 Plumwood, "The Concept of a Cultural Landscape," 117.
3 Bordwell and Thompson, *Film Art: An Introduction*, 82.
4 See Dyer, *Stars*.
5 See Eco, *Travels in Hyperreality*.
6 Burt, *Animals in Film*, 54.
7 Simultaneously, they note the emergence of the representation of animals in the media as threat or danger and this reinforces the belief that they possess bestial characteristics. Gouabault, Dubied and Burton-Jeangros, "Genuine Zoocentrism or Dogged Anthropocentrism?" See also, Armstrong, *What Animals Mean in the Fiction of Modernity*; Baker, *Picturing the Beast*; Philo and Wilbert, *Animal Spaces, Beastly Places*; Daston and Mitman, *Thinking with Animals*.
8 Austin-Smith, "Acting Matters," 19.
9 Michael Kirby drew up a matrix with non-acting at one end of the spectrum and complex acting the other. See Kirby, "On Acting and Not-Acting."
10 Austin-Smith, "Acting Matters," 28.
11 Ibid., 21.
12 Ibid.
13 Peterson, "The Animal Apparatus," 43.
14 Williams, "The Right Horse, the Animal Eye," 30.
15 The difference between the Icelandic horse and other equines is its various gaits. Whereas all horses have four main paces – the walk, trot, canter and gallop – the Icelandic horse has two further paces: the tölt which is a four-beat lateral gait useful for travelling on uneven, rough ground, and the flying pace.
16 Marsh, "Review," 82.
17 Austin-Smith, "Acting Matters," 31.

18 Philo and Wilbert, Animal Spaces, Beastly Places, 5.
19 In Iceland the first law to regulate horse breeding dates from 1891, banning sexually mature colts from running loose.
20 Hastrup, Nature and Policy in Iceland, 254.
21 Burt, Animals in Film, 71.
22 Icelandic horses are good swimmers, their thick coats protecting them from the cold.
23 Plumwood, "The Concept of a Cultured Landscape," 123.
24 Ibid., 117.
25 Hastrup, Nature and Policy in Iceland 1400–1800, 249.
26 Lefebvre, Landscape and Film, 22.
27 Citation from Helen Tiffin, "The Speech of Dumb Beasts," 141.

Bibliography

Armstrong, Philip. *What Animals Mean in the Fiction of Modernity*. Abingdon and New York: Routledge, 2008.

Austin-Smith, Brenda. "Acting Matters: Noting Performance in Three Films." In *Theorizing Film Acting*, edited by Aaron Taylor, 19–32. New York and London: Routledge, 2012.

Baker, Steve. *Picturing the Beast*. Urbana, IL and Chicago: University of Illinois Press, 2001.

Bordwell, David, and Kristin Thompson. *Film Art: An Introduction*. New York: McGraw-Hill Inc., 1993.

Burt, Jonathan. *Animals in Film*. Trowbridge: Reaktion Books, 2002.

Daston, Lorraine, and Greg Mitman, eds. *Thinking with Animals: New Perspectives on Anthropomorphism*. New York: Columbia University Press, 2004.

Dyer, Richard. *Stars*. London: British Film Institute, 1979.

Eco, Umberto. *Travels in Hyperreality*. Translated by William Weaver. London: Picador, 1987.

Fudge, Erica. *Animal*. Trowbridge: Reaktion Books, 2002.

Gouabault, Emmanuel, Annik Dubied, and Claudine Burton-Jeangros. "Genuine Zoocentrism or Dogged Anthropocentrism? On the Personification of Animal Figures in the News." *Humanimalia*, 3 (2011): 77–100.

Hastrup, Kirsten. *Nature and Policy in Iceland 1400–1800: An Anthropological Analysis of History and Mentality*. Oxford, UK: Clarendon Press, 1990.

Kirby, Michael. "On Acting and Not-Acting." In *Acting Re(Considered)*, edited by Philip B. Zarrilli, 40–52. London and New York: Routledge, 1995.

Lefebvre, Martin, ed. *Landscape and Film*. New York and London: Routledge, 2006.

Marsh, Calum. "Review: *Of Horses and Men*." *Sight and Sound*, 24 (2014): 82–3.

Magnússon, Sigurdur Gylfi. *Wasteland with Words*. London: Reaktion Books, 2010.

Möller, Frank, and Samu Pehkonen. *Encountering the North: Cultural Geography, International Relations and Northern Landscapes*. Aldershot and Burlington: Ashgate, 2003.

Nance, Susan. *Entertaining Elephants: Animal Agency and the Business of the American Circus*. Baltimore, MD: John Hopkins University Press, 2013.

Peterson, Michael. "The Animal Apparatus: From a Theory of Animal Acting to an ethics of Animal Acts." *TDR: The Drama Review*, 51 (2007): 33–48.

Philo, Chris, and Chris Wilbert, eds. *Animal Spaces, Beastly Places: New Geographies of Human-Animal Relations*. London and New York: Routledge, 2000.

Pick, Anat. *Creaturely Poetics: Animality and Vulnerability in Literature and Film*. New York: Columbia University Press, 2011.

Plumwood, Val. "The Concept of a Cultural Landscape: Nature, Culture and Agency in the Land." *Ethics and the Environment*, 11 (2006): 116–50.

Sundkvist, Anneli. "Herding Horses: A Model of Prehistoric Horsemanship in Scandinavia – and Elsewhere?" In *PECUS. Man and Animal in Antiquity*. Conference Proceedings at the Swedish Institute in Rome, September 9–12, 2002, (2002): 242–9.

Tiffin, Helen. "The Speech of Dumb Beasts." In *Considering Animals: Contemporary Studies in Human-Animal Relations*, edited by Carol Freeman, Elizabeth Leane and Yvette Watt, 137–52. Aldershot and Burlington: Ashgate, 2011.

Williams, David. "The Right Horse, the Animal Eye – Bartabas and Théâtre Zingaro." *Performance Research*, 5 (2000): 29–40.

3 Horses as Co-Constructors of Knowledge in Contemporary Finnish Equestrian Culture

Nora Schuurman

As a domesticated animal kept for numerous purposes throughout history, the horse can be defined as a hybrid between nature and culture.[1] The practice of keeping horses has radically transformed in the Western world during the past 50 years as there was no need for the horse as a working animal any longer. The common history of humans and horses, however, did not come to an end but opened up a place for the horse in the increasing amount of free time afforded to humans. The changes in human–horse relations include transformations in the knowledge needed to keep horses and care for them. Leisure riding as well as competitive sports have constructed the human–horse relationship in a new way, emphasizing not only the emotional relationship but also the process of learning. The people who keep horses in contemporary contexts commonly have no background in the traditions of keeping the horse as a working animal. Encountering the animal, however, requires knowledge of the animal and the ways of communicating with it and understanding its agency in the daily practices of handling, training, and riding. The question of *knowing* the horse has, therefore, become central in contemporary equestrian culture.

In this chapter, I examine the ways in which the horse, an animal interacting with humans, is understood as a subject and an active agent in the process of knowledge formation. The chapter is based on a study conducted in Finland, where the recent development in human–horse relations serves as an example of the new interest in animals in the Western world. The focus is on the tacit knowledge produced within individual human–horse relationships, and the construction of expertise. I am especially interested in the ways in which the subjectivity and agency of the horse is interpreted, understood, and taken into account in the processes of creating knowledge and expertise concerning horse care and handling. Another issue addressed is the question of animal welfare and how it is understood in relation to the conceptualizations of the horse in contemporary equestrian culture. Caring for the horse in the best possible way has been part of good horsemanship since agrarian times,[2] but in the context of the wider public discussion concerning animal welfare, it is epitomized in specific demands for mental well-being and species-specific behavior, in addition to the more obvious requirement of good health.

Many of the practices of horse care in Finland can be understood as contextual, due to the requirements of and adaptation to the conditions in the north. These include managing the horses in snowy and cold winter weather with stabling and rugs, controlling the impact of biting insects, and operating on the fringe of urban and natural areas, both of which pose specific challenges for horses to move in. In Finland, negotiating the cold winter conditions is an important part of horse management and handling. In the work horse era, it was the horse's responsibility to choose the best route in deep snow,[3] whereas in contemporary equestrian culture, keeping horses warm in stables is contrasted with giving them fresh air and an opportunity to move outdoors in a paddock. Horse keeping is presently an urban activity, since the majority of horse owners live in urban areas and keep their horses at livery yards in peri-urban areas. Finland is a rather sparsely populated country, and the pressure on land use is not as strong as in many other European countries. In consequence, the step to leaving the city to be able to keep horses at home, within a commutable distance of the city, is easily taken.[4] In this chapter, knowledge about horses is explored in the northern context with its special characteristics.

The materials explored for this chapter consist of interviews with leisure horse owners, and internet discussions. I conducted nine thematic interviews with Finnish riding horse owners in eastern Finland in 2007. The interviewees were all female,[5] reflecting the process of feminization in equestrianism common in the Western world.[6] The interviewees were aged between 20 and 60 years, all owners of at least one riding horse, and some of them also had other roles as competitors, riding instructors, trainers, breeders, or yard owners. The horses used for riding in Finland mainly include warmbloods, native Finnhorses, ponies, and Standardbred ex-racehorses. The internet discussions were collected during the years 2008 and 2009 from a Finnish discussion forum about horses, riding, and horse keeping, Hevostalli.net.[7] The forum is the most popular in Finland on the subject, with around 15,000 posts daily at the time of collecting the data. All of the materials were first thematized and then analyzed with a combination of theoretically informed content analysis and discourse analysis.[8] The analysis was based both on descriptions of factual content and on interpreting the meanings and argumentation expressed as discourses of horse keeping within equestrian culture.

Tacit Knowledge and Horses

Keeping animals not only requires knowledge about the animal but also creates it through everyday action and interaction with the animal.[9] Knowledge can be considered as a gateway to animal keeping, and a tool with which the worlds of humans and animals are able to encounter each other and survive in an environment governed by humans. Further, knowledge produced and acquired in various ways and from different sources influences cultural conceptions about animal welfare and its interpretations in everyday contexts.[10] Knowledge about individual animals and ways of taking care of them varies in each circumstance, not least because what is seen as care in some contexts is understood as neglect in others. This knowledge is also influenced by the animals themselves by their own actions.

Instead of theoretical knowledge and predetermined positions, human–animal encounters are primarily based on lived practices and relations,[11] and part of the practical knowledge created in these encounters is not necessarily conscious.[12] Hence, a considerable part of the everyday knowledge about animals can be understood as tacit knowledge. The concept of tacit knowledge refers to the knowledge or skill that is used in action but is difficult to explain verbally.[13] Being practical and embodied, tacit knowledge is acquired through personal experience and is highly contextual. It is often conceptualized as the opposite of explicit or written knowledge, and learning tacit knowledge cannot be reduced to the following of abstract rules.[14] Instead, tacit knowledge can only be learned by observing the work of others and practicing to do the same.[15] Particularly demanding skills are learned in this way, in a traditional master–student relationship, by participating in the work with more experienced actors.[16] A significant proportion of tacit knowledge is also tied to working environments and social practices.[17]

In keeping horses, tacit knowledge has been predominant until the present day.[18] Challenged by the increase in explicit knowledge, especially science and the media, this knowledge has not disappeared but adapted to the new environments and practices of contemporary horse keeping. Tacit knowledge about horses carries information on horse management and care and on ways of communicating and working with them. It can be divided into two categories according to content and type of information: (1) practices such as the use of a harness or stabling, and (2) interaction and communication with horses.[19] The latter is the focus of this chapter, with a special emphasis on the interpretations of the horse as a conscious subject capable of experiencing feelings, emotions, and intentions, and as an active agent whose actions have meaning for the animal itself.

The human–horse relationship is typically based on individual communication and partnership between humans and horses, and this is also the case in contemporary equestrian culture. Focusing on everyday encounters and interactions with animals thus promotes understanding and discussion of the different ways of interpreting animals as individuals and subsequently treating and handling them.

Many individuals such as animal trainers interacting with animals on a daily basis see them as conscious, sentient, communicative, and creative beings.[20] Such accounts are often interpreted as anthropomorphism, a way of thinking that attributes to animals human characteristics, such as thoughts, emotions, motivations, and beliefs.[21] Anthropomorphizing animals has been claimed to be increasing in individual human–animal relationships,[22] and it has been criticized as faulty reasoning, on the grounds of a lack of evidence concerning animal thoughts and alleged welfare risks for the animals themselves.[23] Verbalizing the actions, intentions, and feelings of animals is, however, inevitable in the interpretations of animal behavior in everyday contexts. By observing the animal in the context of the individual human–animal relationship, interpretations taking into account the animal as a conscious subject and agent and sentient being are also possible.

Interaction and Interpretation

According to the analysis of the data, tacit knowledge of the horse is based on everyday interaction with horses, and it is embedded in the embodied practices of caring for, handling, training, and riding them. It is based on personal experience and often described as "an intuitive instinct" that is hard to justify at the time but later turns out to be correct. Knowing the horse as an individual is crucial to be able to understand its subjective actions and experiences, and hence taking lessons at a riding school without further involvement with horses is considered insufficient for acquiring tacit knowledge:[24]

> I didn't have time there [at the riding school] to take care of them and so on, so I just gained more and more experience of riding, while horse care stayed a bit in the background. And then, sometime at secondary school, I stopped going there and went to a smaller yard where I had these moments of clarity: oh, is feeding like this, and so on, and that was where I got my knowledge about horse care and everything. So I think that if I had not gone to the small yard I wouldn't have bought a horse in the first place. Because no, from there, from the riding school, not really, you just learned to put the bridle on, nothing more.[25]

In the data, horse owners reflect their first experiences in learning to know horses individually by interacting with them on a regular basis. They have often been supervised by older, experienced people. Although some of the tacit knowledge passed on from one generation to another can be articulated, there is a lot that cannot.[26] This knowledge typically involves the use of the senses, which requires the teacher to be present, as in the following example: "he taught me hands-on the very basics of horse care and showed me the hay, what good hay was like."[27] The correct appearance, texture, and smell of good quality hay are typical examples of tacit knowledge that is difficult to articulate and express in written form. The individual horse and the specific context also have a crucial significance for constructing knowledge, and many informants see these particular events as turning points in their career as horse owners. This is illustrated by an internet discussant telling about her first years as a horse owner: "now that we have spent more time together I know an awful lot more than when I got the horse."[28] For her, the daily chores of caring for her own horse have become a meaningful experience.

In the individual human–horse relationship, the animal participates in the production of tacit knowledge over time. Despite several years of learning to ride and care for horses, the process of acquiring knowledge based on personal experience is totally different when caring for and handling a familiar horse. During the initial phase, all previously acquired knowledge has to be contextualized in the new relationship, sometimes challenging the previous knowledge. Later, after gaining experience on several different horses, applying the acquired knowledge to new horses becomes easier. A part of the learning process is combining tacit knowledge with explicit knowledge in order to assess the applicability of the latter to the specific context.

In the data, long personal experience is appreciated and used as a justification for arguments for various practices of horse care. This is evident in the following statement, situated in the northern wintry context of using rugs on horses in cold weather: "Myself, I have been involved with riding and trotting horses for about 20 years and still have not met with a horse that would need, for example, three quilted rugs and still no one has been cold."[29] Here, the argument is not connected to explicit knowledge on the ability of horses to withstand cold but, instead, on a long personal experience based on observing the horses individually for a long time. The results of this observation validate the practices used.

In interpretations of human–horse interaction in everyday practices, the horse taking part in the interaction is often described as a conscious subject and an active agent. Such descriptions include feelings and emotions experienced and expressed by the horse, interpretations by the horse of various situations, and intentional actions of the horse. Often the purpose is to empathize with the horse's point of view of the practices that have an impact on its daily life and well-being. In the following, an interviewee ponders on the possible reactions of the horse if kept in the stables: "if he had to stay inside, he would not, he would not stand it."[30]

The process of interacting with and caring for the animal on a daily basis is fairly intensive, and therefore contributes to the construction and transformation of thinking, understanding, and practical knowledge concerning the animal and its care, and the relationship as a whole. Communicating with the animal and interpreting it in various situations is significant. In the next section, I will turn to the skill of "reading" the horse, namely the interpretation of its actions and appearance, as a specific case of tacit knowledge in which the agency of the horse is clearly recognized.

Reading the Horse

In order to understand the agency of the horse, the human has to know the individual animal well enough to be able to interpret its subjective actions and experiences. An integral part of tacit knowledge is the skill of "reading" the horse's embodied communication and welfare.[31] This is based on continuous interaction between the horse and the human and learning to know the individual animal and its life history. In the data, the skill of reading the horse manifests itself as the ability to interpret the horse's emotions, experiences, intentions, and the purpose of its actions. This is done by observing the horse's physical appearance and movement, its actions, and its way of communicating with humans and other animals. The horse may be "happy" or "apathetic," and if it stays in the rear corner of its stable it is "not feeling well." The horse also acts in such a way that the owner has to interpret its messages, indicating, for example, that it wishes to get into the stable or out in the field, or get something to eat. In an example from the internet discussions, the owner concludes that horses standing in the rain while there is a shelter available are making a conscious choice because they are not feeling cold and hence there is no need to use rugs on them to keep them warm. In another example, the horse stops drinking when feeling stressed. The data illustrates the effort required on the part of the human to recognize the agency of the horse in their shared everyday life. Only a thorough knowing of the horse and its daily actions enables the owner to interpret the horse's messages and subtle changes in the horses' actions and react accordingly.

Horses also seem to appreciate the attentiveness of the human. Experienced horse owners, skilled at reading their horses, describe their horses as being prone to giving feedback, as stated by the following interviewee: "the more you ask them, the more they give answers. And they do have quite a few opinions."[32] The feedback given by a horse is often regarded as a criterion of correct knowledge concerning methods of horse care and training. This feedback, similar to what Holloway describes as farm animals "acting back" on farming practices,[33] is given by horses to influence the numerous daily decisions concerning their own life. In the data it is given so much weight that the informants are ready to change their practices on the basis of this feedback. Horses themselves also read humans when interacting with them, and in such instances the animal may sometimes turn out to be the more knowledgeable one. In the work horse era, for example, draft horses often knew their work and the specific tasks they had to perform in each phase of work.[34] They thus shared the responsibility for the brainwork that had previously been attributed to humans only.[35]

The skill of reading the horse develops as a process of mutual learning where the human learns to read the horse while the horse learns to communicate its feelings, needs, and intentions to the human in a way that the human can understand. In the first relationship with a horse, the learning process takes time and the communication only works after the horse becomes familiar enough for the owner to learn how to read its messages. This takes place in concrete, daily practices and situations experienced with the horse, and the relationship becomes the essential context for the owner's tacit knowledge:

> I suppose I can say by now that I clearly see if, ok, she has a bit of a bad day or, ok, now she has a good day, she is such a personality, she is, you can see it clearly … I probably would not, in the beginning, I would not have seen all the little things, when I did not know how she normally behaved.[36]

In a livery yard, equine welfare becomes a social issue among the community of horse owners. The horse's appearance may be overemphasized as it is important for the owners that the horse not only feels well but also looks as if it is feeling well. Instead, interpreting the horse's subjective actions and experiences and communicating with it feels easier when the horse is kept at the owner's stables at home, with the possibility of observing the horse's behavior and appearance on a daily basis. As living and keeping a horse in this way is within reach for ordinary horse owners due to reasonable real estate prices, such solutions have become rather popular. They are also beneficial for the human–horse relationship:

> In livery, there is the downside that you do not know. You don't know in the morning, during the day, in the evening, how your horse is doing. I know

> immediately if, ok, everything is not normal, if the muck is in a different place or if there is uneaten hay or … I see from the stables, too, how they are doing. So many things tell me about the horses' well-being.[37]

Again, the importance of reading the horse's subjective actions can be illustrated with examples of horse care from wintry contexts. While practices such as stabling are issues of much controversy within discussions among horse owners,[38] an owner who is experienced enough to read her horses makes the decision according to her interpretation of the horses' messages: "if they keep standing there at the gate, looking like, no, it is terrible out here, I tell them to come in."[39] In the communication with the horse, the horse is given an active role in the dialogue:

> Sometimes, when the weather is really lovely and I go out to get them in at half past nine in the evening and the horses say, uh, we aren't coming, they can stay out for the night in the wintertime too, it's no problem for me or them … and my horses can speak, I can ask them, will you come in, oh, you won't, well do you want food outside, ah, you do. Okay, stay out for the night.[40]

As a result, the owner has learned to know the kind of weather her horses prefer for staying out – not necessarily the warm nights but the dry and still ones. In such cases, the horse is understood as a subject and an agent making choices concerning its own everyday life, and feedback given by the horse influences the practices of its care. As one of the interviewees states, "life with horses becomes a lot easier … when you have learned to read the answer from the horse and you do not have to wonder why this horse acts up."[41] In this sense, the animal actively participates in the construction of knowledge of itself and ultimately of its own welfare.

The Horse Constructing Expertise

When described as an intentional agent, the horse sometimes expresses its views so actively that it actually takes charge of the situation. This happened to one of the interviewees while riding in a competition: "I remember how I thought, there in the arena, that those jumps were horrible and [the horse] was like, yes Mom, we'll make this."[42] The horse is also considered to possess knowledge and to make conclusions about matters concerning its own daily life, as in the following: "We had this old mare, she did not even eat sugar. She kind of took it in her mouth and dropped it out. She probably knew that it wasn't good for her."[43] In such situations, the animal is understood to be an expert of its own life. Interestingly, the horse may also make wrong decisions, an aspect of subjectivity and agency that may be difficult for some owners to accept. Others, however, see this as an essential part of the horse's learning process. Serious mistakes can, nevertheless, result in open conflicts and accidents endangering the whole relationship and the safety of both human and horse, which is why the human always controls the agency of the horse to some extent, while the role of the horse is one of submission and obedience.[44] A human–animal relationship "inevitably involves some forms of restriction, power and control,"[45] and this also applies to the human–horse relationship despite its ultimate aim, a partnership with mutual communication, trust, and respect. The more successful the relationship becomes, the more subtle is the exercise of human power.

The agency of the horse also has implications for the construction of expertise in the contexts of horse care and handling. The feedback given by the horse is used to assess the expertise of various professionals such as yard owners, farriers, and trainers, who have a focal role in the newly popularized and commercialized equestrian culture. Understood as socially constructed, expertise is always constituted in interaction with a specific audience or a community that recognizes the skills or knowledge of the expert. Thus, expertise cannot only be defined as a profession based on formal qualifications, but as a social status, with the line between an expert and a layperson drawn contextually in each situation.[46] A person aiming to achieve the status of an expert has to be able to prove the superiority of his or her expertise compared to other expertise and to construct others as laypersons. Choosing the appropriate experts is a challenge for horse owners, especially for those with less personal experience and hence the greatest need for support. The expertise of various service providers is assessed by horse owners by observing their way of talking about horses and working with them in actual encounters. Feedback from other horse owners is also considered to be an indicator of the expert's trustworthiness.

Apart from fellow humans, horses themselves also participate in the construction of expertise in horse keeping. In this process, the meaning of the subjective action by the animal is constructed by the human.[47] An essential part of constructing expertise is interpreting the horse's reactions to the actions of the expert. A visibly contented horse is likely to indicate the expertise of the service provider, whereas a horse that unexpectedly refuses to cooperate is understood to signal the opposite. Fox example, horses may not trust veterinarians who do not know how to handle horses, or they may lose their trust in farriers who they perceive as aggressive: "He started shouting and hitting and worse, and that was when I changed the farrier because I knew that she could not have trusted him anymore, and just by seeing him, she would have jumped all over the place and run away."[48]

A central question here is trust. Trust is a prerequisite for the acknowledgment of expertise, and trust in experts is often based on the knowledge of trust experienced by others.[49] Trust can also be lost if something happens that raises doubts about the competence and credibility of the expert. Similarly, a horse may not trust a certain trainer whose ways of working seem unreliable or illogical. In conflict situations, the feedback given by the horse may be valued more highly than human expertise, as in the case of an interviewee training showjumping. Here, the horse is understood to know better:

> I have trained with X for the past fifteen years, and his riding style is very soft, very considerate for the horse, with the idea that the horse is very intelligent and can find the stride on her own. And Y's style is such that the horse does not understand that the rider has to find the stride for her, that they are

> so stupid that they don't know where they take off. So I had to have a little conversation about this with him ... of course my horses had been taught to look after themselves.[50]

The mistrust between horse, rider, and trainer led to an accident, after which the rider decided to trust the horse rather than the trainer. The example also illustrates the significance of context, or the absence of context, in tacit knowledge. The trainer coming from outside is not knowledgeable about the specific ways of communicating and reading each other in the particular rider–horse relationship.

In the data, the horse is sometimes described as a teacher, and therefore an expert itself on human–horse interactions. Learning from a horse is embodied and based on individual communication. A very experienced riding horse is called a schoolmaster, referring to the master–student relationship central in learning tacit knowledge.[51] Such a horse is able to assist its rider in developing her riding skills, which is why these horses are commonly bought for young and fairly inexperienced riders. Long relationships with experienced schoolmasters are highly valued: "I could keep him as if he was my own, and with him I actually learned all I know about riding."[52] Schoolmasters are appreciated partly because they act in the way they are taught to and thus reflect the system constructed by humans, and partly because they ultimately act independently, correcting any mistakes made by the inexperienced rider. The expertise of the horse is often considered unbiased and therefore more important than that of any given trainer, and riders are advised to always listen to their horses for feedback. Appreciating the horse as an expert is ultimately expressed by the interviewee who states that horses "know all about horses, absolutely. We will never learn to know as much about horses as they do."[53]

Conclusion

In this chapter, I have discussed the active role of the horse as a co-constructor of knowledge concerning its own everyday life and well-being. The role of the horse in producing knowledge is based on tacit knowledge of the horse as an individual, acquired through interaction with the horse in the various practices of care and handling on a daily basis. This knowledge develops over time, and requires both the opportunity and the skill to read the horse's appearance, messages, and feedback on human actions. Significant to this process is the interpretation of the horse as an individual agent instead of as a mechanical recipient of routine care. The central role is played by the horse itself, as much as by the person reading the horse and making the interpretations. Does interpreting the horse verbally equal to anthropomorphizing it? I would not say so. Rather, it seems that there is a place in equestrianism for the horse as an individual, a conscious subject, and an intentional agent, albeit controlled, within the cultural interpretations of the horse as a hybrid between nature and culture.[54]

The study illustrates, however, that human–horse interaction, including the appreciation of the agency of the horse, is essentially situated. Understanding knowledge about horses as being primarily based on personal experience in the context of an individual human–animal relationship makes visible the significance of the daily practices in caring for and handling horses. When decisions are made on the basis of tacit knowledge and intuition, they are made in the specific context of the individual actors, the yard, and the actual situation at hand. Although general explicit knowledge is used it is not applied as such but always in relation to contextually produced tacit knowledge. As in the examples discussed in this chapter, the northern context, epitomized in wintry conditions or land use, bears its own consequences on the practices of horse keeping. These practices, their justification, and their significance in different situations must be understood by both human and horse, in the particular environment that they live in.

The horse is also given a role in the relations between humans, as is exemplified in the construction and definition of expertise. The agency of the horse is especially significant in creating the trust that expertise is based on: the horse is included in the social process in which the status as an expert is formed. The horse trusts or does not trust the expert, and the owner then trusts the horse's judgment. The owner's interpretations of the horse become central criteria for expertise, constructing a role for the horse as an expert on its own life. In this way, the horse's voice is heard in matters concerning itself and it is able to influence its own welfare, as Buller and Morris suggest.[55] Furthermore, by participating in the construction of human expertise, the horse is understood as part of the (human) social world, and its actions also have consequences for humans. What is essential for this process is the interaction between owner and horse, supported by the owner's ability to interpret her horse as an individual animal, a conscious subject, and an intentional agent.

Notes

1 Greene, *Horses at Work.*
2 Leinonen, *Palvelijasta terapeutiksi.*
3 Ibid., 120–1.
4 Elgåker et al., "Horse Keeping in Urban and Peri-Urban Areas."
5 In Finland, leisure riding is predominantly a female activity, as more than 90 percent of those who ride are women and girls. The Equestrian Federation of Finland, "Facts and Figures."
6 Adelman and Knijnik, "Introduction."
7 *Hevostalli* is Finnish for horse stable.
8 Tuomi and Sarajärvi, *Laadullinen tutkimus*; Potter, "Discourse Analysis."
9 Ingold, *The Perception of the Environment.*
10 Buller and Morris, "Farm Animal Welfare."
11 Irvine, *If You Tame Me.*
12 Ingold, *The Perception of the Environment*, 52.
13 Polanyi, *The Tacit Dimension.*
14 Toom, "Hiljaista tietoa vai tietämistä?"
15 Polanyi, *The Tacit Dimension*, 30.
16 Toom, "Hiljaista tietoa vai tietämistä?"
17 Onnismaa, "Hiljainen tieto kulttuurien rakenteissa."
18 McShane and Tarr, *The Horse in the City.*

19 Greene, *Horses at Work*.
20 Arluke and Sanders, *Regarding Animals*.
21 Serpell, "Anthropomorphism and Anthropomorphic Selection."
22 Franklin, *Animals and Modern Cultures*, 97.
23 Serpell, "Anthropomorphism and Anthropomorphic Selection."
24 Birke, "Talking about Horses."
25 Interviewee 22 years. No reference to individual interviews other than age at the time of interviewing is used since the interviewees were promised anonymity.
26 Polanyi, *The Tacit Dimension*; Toom, "Hiljaista tietoa vai tietämistä?"
27 Interviewee 44 years.
28 Hevostalli.net, "Foorumi," April 4, 2009.
29 Hevostalli.net, "Foorumi," November 11, 2009.
30 Interviewee 20 years.
31 Birke, "Talking about Horses."
32 Interviewee 40 years.
33 Holloway, "Pets and Protein."
34 Greene, *Horses at Work*, 270.
35 McShane and Tarr, *The Horse in the City*, 175–6.
36 Interviewee 22 years.
37 Interviewee 40 years.
38 Schuurman, "Conceptions of Equine Welfare."
39 Interviewee 33 years.
40 Interviewee 40 years.
41 Interviewee 60 years.
42 Interviewee 40 years.
43 Interviewee 22 years.
44 Thompson, "Theorising Rider–Horse Relations."
45 Fox, "Animal Behaviours, Post-Human Lives."
46 Arnoldi, "Universities and the Public Recognition."
47 Crist, *Images of Animals*.
48 Interviewee 22 years.
49 Wynne, "May the Sheep Safely Graze?"
50 Interviewee 44 years.
51 Toom, "Hiljaista tietoa vai tietämistä?"
52 Interviewee 23 years.
53 Interviewee 60 years.
54 Greene, *Horses at Work*.
55 Buller and Morris, "Farm Animal Welfare."

Bibliography

Research Materials

All prints of the internet discussions, audio, and transcripts of the interviews are in the possession of the author.

Literature

Adelman, Miriam, and Jorge Knijnik. "Introduction – Women, Men, and Horses: Looking at the Equestrian World through a 'Gender Lens.'" In *Gender and Equestrian Sport*, edited by Miriam Adelman and Jorge Knijnik, 1–14. Dordrecht, Netherlands: Springer, 2013.

Arluke, Arnold, and Clinton R. Sanders. *Regarding Animals*. Philadelphia: Temple University Press, 1996.

Arnoldi, Jakob. "Universities and the Public Recognition of Expertise." *Minerva* 45 (2007): 49–61.

Birke, Lynda. "Talking about Horses: Control and Freedom in the World of 'Natural Horsemanship.'" *Society and Animals* 16 (2008): 107–26.

Buller, Henry, and Carol Morris. "Farm Animal Welfare: A New Repertoire of Nature-Society Relations or Modernism Re-embedded?" *Sociologia Ruralis* 43 (2003): 216–37.

Crist, Eileen. *Images of Animals. Anthropomorphism and Animal Mind.* Philadelphia: Temple University Press, 1999.

Elgåker, Hanna, Stefan Pinzke, Gunilla Lindholm, and Christer Nilsson. "Horse Keeping in Urban and Peri-Urban Areas." *Geografisk Tidskrift – Danish Journal of Geography* 110 (2010): 81–98.

The Equestrian Federation of Finland. "Facts and Figures: Equestrianism in Finland." Accessed October 2, 2014. www.ratsastus.fi/.

Fox, Rebekah. "Animal Behaviours, Post-Human Lives: Everyday Negotiations of the Animal-Human Divide in Pet-Keeping." *Social and Cultural Geography* 7 (2006): 525–37.

Franklin, Adrian. *Animals and Modern Cultures.* London: Sage, 1999.

Greene, Ann Norton. *Horses at Work. Harnessing Power in Industrial America.* Cambridge, MA: Harvard University Press, 2008.

Holloway, Lewis. "Pets and Protein: Placing Domestic Livestock on Hobby-Farms in England and Wales." *Journal of Rural Studies* 17 (2001): 293–307.

Ingold, Tim. *The Perception of the Environment. Essays in Livelihood, Dwelling and Skill.* London and New York: Routledge, 2000.

Irvine, Leslie. *If You Tame Me: Understanding Our Connection with Animals.* Philadelphia: Temple University Press, 2004.

Leinonen, Riitta-Marja. *Palvelijasta terapeutiksi: ihmisen ja hevosen suhteen muuttuvat kulttuuriset mallit Suomessa.* Oulu, Finland: University of Oulu, 2013.

McShane, Clay, and Joel A. Tarr. *The Horse in the City. Living Machines in the Nineteenth Century.* Baltimore: The Johns Hopkins University Press, 2007.

Onnismaa, Jussi. "Hiljainen tieto kulttuurien rakenteissa." In *Hiljainen tieto: tietämistä, toimimista, taitavuutta*, edited by Auli Toom, Jussi Onnismaa and Anneli Kajanto, 83–102. Helsinki: Kansanvalistusseura and Aikuiskasvatuksen Tutkimusseura, 2008.

Polanyi, Michael. *The Tacit Dimension.* Gloucester, MA: Peter Smith, (1966) 1983.

Potter, Jonathan. "Discourse Analysis." In *Handbook of Data Analysis*, edited by Alan Bryman and Melissa A. Hardy, 607–24. London: Sage, 2009.

Schuurman, Nora. "Conceptions of Equine Welfare in Finnish Horse Magazines." *Society and Animals* 23 (2015): 250–68.

Serpell, James. "Anthropomorphism and Anthropomorphic Selection – Beyond the 'Cute Response.'" *Society and Animals* 11 (2003): 83–100.

Thompson, Kirrilly. "Theorising Rider-Horse Relations: An Ethnographic Illustration of the Centaur Metaphor in the Spanish Bullfight." In *Theorizing Animals: Re-thinking Humanimal Relations*, edited by Nik Taylor and Tania Signal, 221–53. Leiden, Netherlands: Brill, 2011.

Toom, Auli. "Hiljaista tietoa vai tietämistä?" In *Hiljainen tieto: tietämistä, toimimista, taitavuutta*, edited by Auli Toom, Jussi Onnismaa and Anneli Kajanto, 33–58. Helsinki: Kansanvalistusseura and Aikuiskasvatuksen Tutkimusseura, 2008.

Tuomi, Jouni, and Anneli Sarajärvi. *Laadullinen tutkimus ja sisällönanalyysi.* Helsinki: Tammi, 2002.

Wynne, Brian. "May the Sheep Safely Graze? A Reflexive View of the Expert-Lay Knowledge Divide." In *Risk, Environment and Modernity: Towards a New Ecology*, edited by Scott Lash, Bronislaw Szerszynski and Brian Wynne, 44–83. London: Sage, 1996.

4 Living with Horses

Horse Agency in Human–Horse Cohabitation in Nineteenth-Century Finnish Swidden Culture

Riitta-Marja Leinonen

Introduction

The relationships between humans and animals in the North is currently a popular topic in anthropological research; however, the focus is often placed on nomadic peoples, hunters, and fishermen.[1] The lives of farmers in the boreal forest region, on the other hand, have received little attention from anthropologists, with the challenges of farming and keeping animals in the North remaining more in the realm of interest of folklorists,[2] ethnologists,[3] and archeologists.[4] Furthermore, anthropological studies have been criticized for neglecting the agency of animals and presenting a dualistic and overly simplistic view of farmers as having an instrumental, anthropocentric, and hierarchical relationship with animals and nature.[5] Opposition to this view can be found in another over-generalized perspective, a highly idealized one in which pre-modern, non-Western small communities exist in an equal and reciprocal relationship with animals and nature.[6] Anthropologist Dimitros Theodossopoulos argues that human–animal relationships in small rural communities should be investigated context-specifically, with attention paid to care, reciprocity, and understanding of the place and position of each living being on the farm.[7]

This chapter focuses on the human–horse relationship in the context of cohabitation in rural settings of central and eastern Finland in the nineteenth century from the perspective of cultural anthropology and animal agency. Cohabitation between humans and farm animals is known to have existed around the world, and indeed it still does. As archeologist Kristin Armstrong Oma has stated, built environments were "made with particular reference to human–animal relationships."[8] In Finland, humans and animals shared the same room in wintertime. Here I examine this co-living of humans and horses in shared life space.

The human agents considered in this chapter are nineteenth-century farmers in central and eastern Finland who led self-sufficient lives engaged in a diverse economy, with slash-and-burn cultivation, keeping of cattle, hunting, fishing, and gathering. Ecologically, geopolitically, and culturally Finland is situated between East and West,[9] and a natural boundary formed by river valleys and eastern hills separates it into flatlands and the hill country. In the nineteenth century, these also marked the boundary between swidden cultivation and cultivation by plough, demarcated the eastern and western dialects, and delineated differences in other cultural elements such as material culture and spiritual heritage.[10]

The non-human agents I will consider are native Finnish horses, most likely the eastern type called the small Karelian horse by hippologists Gummerus and Alfthan (see Figure 4.1). These animals were about 140 centimeters high but varied in size and looks.[11] They were described as hardy, strong, stubborn, and economical to feed.[12] In the slash-and-burn areas, horses were used for harrowing and ploughing and for traveling the roadless terrain.

I start by presenting the research material, then delve into the central themes of this chapter: the life and worldview of swidden cultivators; the practices of horse-keeping in the nineteenth century; horse agency in that context; and, finally, the changing horse-related concepts in developing urban and rural mindsets.

The Research Material

My main research material of human–horse cohabitation comes from ethnographic texts and folklore from the Folklore Archives of the Finnish Literature Society. From the ethnographic corpus I have constituted my data consisting of texts by 15 people who interviewed rural women and men or wrote about their observations between 1893 and 1960. The interviewers were usually educated people of the day, such as policemen and university graduates, and often collected information in their home municipality. Some of them traveled around, pursuing a hobby of folklore collection. They wrote about how things happened "in the olden

Figure 4.1 A small Finnhorse stallion Tähti (Star) in Ilomantsi, Eastern Finland, 1910.

Source: Photo by A.G. Nissinen, Ilomantsi. Photograph Archives of the Museum Foundation of Ilomantsi.

days," which, while naturally a very indefinite way of describing time, usually referred to the setting of stories heard from informants' parents and grandparents. Hence, in this material, it refers to the nineteenth century. Four of the interviewers had spoken with several people. Interviewees were 23 to 89 years of age and were identified by name and categorised farmer, farmer's wife, granny, or old man of the croft. For a fuller picture of the worldview and human–horse relationship of the swidden cultivators, I have also examined folklore material, mainly spells and magic rites, from the Old Poems of the Finnish People collection.[13]

The ethnographic data are fragmented and not very large in number, but they can be usefully read in parallel with previous research, from the 1970s and 1980s, by ethnologists Asko Vilkuna and Maija-Liisa Heikinmäki, whose research was based on ethnographic fieldwork and material from the archives of the Finnish Literature Society and National Board of Antiquities.[14] The discussion here is partly based on the data they used, but it complements the ethnologists' perspective with the focus of current anthropological discussion surrounding human–animal relationships and animal agency. In the ethnology tradition, the animal's agency is rarely seen, at least in older ethnographic texts, which concentrate on describing local customs and material culture at a general level. To find the agency of the horse in the material, I performed a sensitive reading and "read the animal" in the text.[15] I did this by using the anthropological theories of human–animal relationship, current scientific knowledge on horses, and the history and ethnography of northern farmers. My previous research, including fieldwork on the human–horse relationship in Finland, alongside my experiences with horses, also had an effect on how I interpreted the research material. One of my family members is a warmblood riding horse, and I have known several Finnhorses, who are the stubborn descendants of the resilient native horses in the earlier farmers' lives.

I asked my research material why horses were brought inside the house in winter. More broadly, I considered how the northern climate and the culture of these farmers of the North, especially their worldview, affected their horse-keeping. How was the cohabitation of horses and humans described in the texts, and by whom? What was horse agency like? How did living together affect the human–horse relationship?

The Horse in the Worldview of Swidden Cultivators

Natural environment, living conditions, and day-to-day subsistence affect people's worldview and how they interpret their surroundings and fellow beings.[16] Life was hard in the nineteenth-century northern setting for all, humans and domestic animals alike, with scarce or poor-quality food, hard work, and vermin.[17] Winters were long, with snow covering the ground for five to six months a year, and summers were short, so people had to practice diversity in their economy, not relying on farming alone.

The swidden cultivators lived in remote villages in the boreal forests. This helped to preserve a worldview that was a combination of local ancient religion and Christianity. Even though Christianity had been introduced about 500 years earlier in that region, the farmers still relied on magic to secure their future and to safeguard everyday life. Ritual tradition with spells was a way to control reality, and, for example, people integrated Christian influences into those spells.

People believed that there was a spirit in every living being and dwelling. Places, animals, people, and elements such as water and wind had a guardian spirit (*haltija*). *Haltijas* were the keepers of happiness and prosperity of a house and protectors from threats. If someone made the *haltija* unhappy, the spirit would leave and bad luck would follow the people and animals of the house.[18] Horses were to be treated with respect because mistreating a horse was avenged by either the horse or the horse's *haltija*.[19]

The native horses were seen as wise animals with a good memory. They were considered capable of having strong feelings and of having a strong attachment to their family. The special connection between a man and a horse was emphasized by a belief that the horse would know if its master died even when in another physical location.[20] The horse was believed to be attached to its home place, to be *kotipaikkauskollinen* (literally "loyal to his or her home place"), meaning that she had the ability to become attached to the people there and to be held there by the *haltija* or the spirit of that place. This influenced the practice of selling horses; if they were sold, they were sold young. When buying a horse, one had to shake hands, in a sign to the earth spirit that the horse was leaving the place.[21] If one bought an adult horse, it was convenient to know how to use a spell to prevent that horse from running away. People feared also that the horse would miss her previous home so much that she would stop eating and lose her will to live.[22]

In this context, the horse was seen as not just a workmate and a household member but also as part of the other world. The borders between the world of the living and that of spirits and the dead were porous and overlapping. The ancestor cult was a central part of the religious life of these people, and burial grounds were visited regularly, especially in the areas under the influence of the Orthodox Church. According to folk knowledge, these burial grounds, or sacred groves (originally called *hiisi*), were the horse's origin. This was important knowledge, for, in order to control any phenomenon, one needed to know its birth, its origins.

The strength of the horse was considered to be both physical and mental: these animals were thought to be able to carry people's worries. Even though the therapeutic significance of the horse was not officially acknowledged until the middle of the twentieth century, folklore presents the horse as carrying the burden of people's worries – for example, in the following extract from an old poem:

> Let the horse worry,
> The horse has a bone firmer,
> Bone firmer, head harder.[23]

Humans, animals, places, spirits, and Christianity formed a consistent reality in the world of swidden cultivators, where everything was alive and all living things were connected with each other, also including evil elements from which one needed protection. This worldview formed the ontology that affected the day-to-day life of the farmers and their non-human companions.

Horses Living inside the House

The practice of humans and animals sharing the house in winter was typical in eastern and central Finland.[24] The time horses spent inside depended on the weather and the need for care. There were no actual stables, just a draughty place called *kuja*, or alley, between two buildings. The stable was of western origin, and it was the last of the farm outbuildings to reach eastern Finland, being introduced at the dawn of the twentieth century.[25] The house was the only warm place and was home also to other animals – such as hens, piglets, calves, and lambs – in cold weather.[26] The shared life space facilitated caring for horses and other livestock.[27] Horses ate, drank, and were shoed, groomed, and even washed inside. They were also kept indoors if they were ill or if people wanted to train them or just pet them.[28]

The horse's place in the house was at the side of the door, where the horse trough (*hevosamme*) was situated (see Figure 4.2). There was also a large metal ring in the wall, to which the horse was attached during shoeing and while eating, although in some houses horses remained loose. The horse was brought inside particularly to be taken care of, to dry and warm up in the winter, and to eat moist mash that would have frozen in a cold outbuilding.[29] This mash, an important component of the diet alongside hay, was made from straw, oat flour, and water. Additionally, fresh water was added in the trough by the door each day.[30]

> In the olden days, horses were always fed and watered inside the house. They were not groomed, for they did it themselves by rolling in the snow and swimming in the summer. … The drinking trough was always full. Rowan water was boiled especially in the spring; it protected the horse from disease.[31]

The above is an example account written down in 1947 by Osmo Korosuo, a university graduate, who interviewed Toivo Haukivaara, a 24-year-old farmer in Taivalkoski. The text is not very detailed and leaves many questions, such as what disease rowan water protected a horse from.

Shoeing horses became more commonplace by the end of the century, and the animals were shoed inside the house in wintertime because this was the only warm place with sufficient light. It was said that horses were calmer to shoe inside, probably because it was a familiar and safe place where they were fed and able to warm up.[32] This is interesting because the ethnographic texts usually described shoeing a horse as a rather violent activity; accordingly, there were spells for keeping the horse calm during shoeing.[33] Possibly, people were calmer inside the house and this had an effect on horses. Furthermore, the research material mentions that people spent time petting the horse, and this must have added to the positive feelings associated with staying inside.[34]

Figure 4.2 A Finnhorse eating mash inside a house in Ristijärvi, Eastern Finland, 1932.

Source: Photo by Ilmari Manninen. National Board of Antiquities, Folklore photo collection.

According to the ethnographic texts, the horses were house-trained and just as tidy in the house as the dogs and cats inhabiting it. They left the building when they needed to relieve themselves. If they were loose, they just pushed the door open, and a horse that was haltered and tied caught the family's attention with a slight stamp of a foreleg. Otherwise, people had to keep an eye on the horse to see whether it "gave the sign with its tail"; in other words, the other residents had to know how to interpret the horse's signals.[35] Swedish officer K. F. Uggla wrote about the matter when describing circumstances in the northern Savo region in 1790:

> The horses are more spirited than people, because when they want freedom, they push the door open themselves, go out, come back, and on their return stamp on the stairs in order to inform of their arrival. After having eaten enough mash and hay, they will lie down like dogs at the center of the floor. They are never shod, and they never kick but require fierce teaching before the young horses learn to take care of the required tidiness.[36]

The reference to tidiness means that the horse had to learn to urinate and defecate outside. In one source, the house-training of horses was said to be performed with whip and roar.[37] One cannot generalize from this – as always, there were people who used violent and negative training methods and people who were consistent and gentle with animals. The treatment of animals was also subject to the gendered practices of the era. Horses were mainly the work partners of men, and showing affection was not part of the male role at that time.[38]

The small native horses of the swidden cultivators were not totally dominated by humans; they were free to breed and graze as they wished in summer pastures, organizing in social groups of their own, and when living with humans part-time in the winter, they were still able to go in and out as they pleased. Finnish hippologists called the horse-breeding in these circumstances natural breeding (*luonnonkasvatus*).[39] Such human–animal relationships were based on companionship and trust, and they can be characterized as a social contract that gives the animal shelter from the cold and predators, along with food to get through winter, and to the human, on the other hand, the cooperation and strength of the animal in their work.[40]

Horses' Agency

The idea of horse agency is twofold in the sense that it is something horses possess as intelligent, social animals and, on the other hand, something that can be allowed for them by humans, depending on human ideas of horses' abilities. Horses are actors with their own mind, history, experiences, emotions, morals, and manner of reasoning.[41] There is abundant scientific evidence of animals' cognitive abilities and complex behavior.[42] For example, horses brought up in human societies are intent on staying with people they like and communicating with them. They co-construct the world they share with humans, and at the same time they act on their specific forms of agency, creating their own worlds. Their world is not separate from humans', especially when sharing living space.[43]

Agency can be said to rest on action. Horse agency in the context of nineteenth-century swidden culture rested on interaction with other domestic animals and humans in wintertime and with other horses and life in the woods in the summer. Horses were united with humans by common experience, training, habituation, and sharing of the same life space. People counted on them and their abilities.[44] The agency of a horse was valued, and both horses' cognitive skills and their capacity to learn were appreciated. People expected horses to go out as necessary, to find their way back home from their travels even though their driver might be sleeping, and to find food in the summer pastures.

Living closely with humans changes an animal, and vice versa. Domestication is seen today as co-evolution, taken to entail a symbiotic and dynamic relationship between humans and animals. Living together confers both an opportunity to observe each other and a chance to engage in close interaction. Humans and non-human animals are entangled with each other's life at microbiological and ontological levels as well. We are part horse, cow, sheep, hen, pig, cat, and dog in microbiological terms and also culturally.[45] Zoonotic diseases – diseases that can be transmitted directly or indirectly between animals and humans – are one example of this.[46]

Social animals – horses and humans among them – living in complex social groups develop shared social, cognitive, and communication skills.[47] For example, humans and horses are known to form a language of their own in interspecies communications.[48] When living together, humans and horses learn to interpret each other. They become attuned to each other as they spend time together. Just as mutual grooming makes the bond between horses stronger and relaxes them,[49] the equivalent occurs in interspecies relationships such as that between a human and a horse. Armstrong Oma describes this kind of relationship as "resilient, complex, conscious, reciprocal, close, sensory, tactile and affined."[50]

Trust is a central feature of any human–horse relationship. In the swidden-culture households, the trust between the small native horses and their human companions grew out of the practices of everyday life and with familiar rhythms of days, seasons, and years.[51] Humans trusted the horse to be docile and to work for them, and the horse trusted humans for food and protection. Gaining mutual trust demands time and training. As Kari Weil states, through training, the animal can be freed from the position of victim, because human and animal cultures meet in training and a common language is built with vocal and gesture signs.[52] One can well ask how well trust was achieved when the training methods were violent. If training was successful, and trust achieved, it formed "the basis for animals to give up their autonomy and for humans to claim responsibility."[53]

Horses as socially skilled animals brought warmth and acceptance to people's lives. They needed people and showed affection.[54] The close relationship is acknowledged in the ethnographic text of K.E. Kallio and also the research done by Vilkuna. These authors considered the animals to be well-kept because they lived as humans did and there was a strong bond between an animal and a person who took care of that animal. According to Vilkuna, this was visible even as recently as the 1950s in the way that the small-scale farmers were attached to their cattle and horses. Kallio was especially affected by a belief in the "supernatural senses" of the farm animals and the "poetic and sensitive" relationship towards animals. He was impressed by the way in which the animals trustingly approached strangers and felt safe around them, writing that, in fact, animals were more skilled in interpreting human feelings than were humans themselves. Nevertheless, he was worried about the mechanization, haste, and noise ushered in with the era of modernization, which he was concerned might bring alienation into "the understanding and safe relationship between humans and animals."[55]

Changing Conceptions about Domestic Animals at the Turn of the Century

In the era of modernization, horses were the most common means of transport. Therefore, most people had to handle them, even those who were not interested in their well-being. It is not a coincidence that ideas of animal protection started to spread in the late nineteenth century, when the use of horses reached its peak both in the cities and in the countryside. Reasons for the animal-protection movement lay in city people's divergent conceptions of animals relative to rural people's. As Ovar Löfgren states, the relationship of the upper class and city people to animals was affected by scientific knowledge and a growing distance from animals and nature, which cast more exotic and romantic tones over the relationship to them. A second reason was class-based thinking wherein the upper class considered themselves good and civilised animal-keepers, while superstitious farmers and working-class people were the ones who beat their horses and neglected their dogs.[56]

Ideas and practices related to cleanliness differed between the cities and the countryside, where cattle manure was not dirt but a sign of wealth. Even dirt itself was not dangerous there, and neither was the closeness of the animals; ill wishes and losing cattle-luck, however, were. The smell in peasants' houses with animals in them was, beyond doubt, different from the aromas of upper-class city homes. For people in cities and on wealthier farms, cohabitation between humans and animals was repugnant not only because of the smell but also on account of the primitive living conditions and the mixing of human and animal spaces. Class differences are crystallized, for example, in a story told by a city-dweller who described the smell and scene in a small house in central Finland where the poor inhabitants and their animals were "lying here and there" on the floor.[57] Folk ideas of cleanliness were interpreted as a sign of moral decay.[58]

Peasant ideas of magic and cattle-luck were slowly replaced by new cultural values and other conceptions of animals. As literacy and education grew more common, scientific knowledge about the well-being of domestic animals became known also in remote villages, but that was not until the beginning of the twentieth century. In the cities, ideas about animals started to change as pets became more popular. In towns, dogs, cats, small birds, and goldfish were the only animals allowed in the living room.[59] In remote villages in eastern Finland, the introduction of the stable removed horses from the house, placing them in their own sphere, and the tradition of human–horse cohabitation came to an end. This can be interpreted as reverse domestication, wherein the distance between humans and non-humans grows.[60]

Conclusion

Humans and other animals have lived and evolved together for millennia, and agricultural life in the North would not have been possible without the local domesticated breeds. Human–animal joint communities have produced food together and shared it, and they have shaped their environment in concert. Native breeds and humans have co-evolved and adjusted to the environment, creating new cultural environments that can be called traditional landscapes.[61] While humans have domesticated animals, we too have been domesticated in the process, by animals.[62]

The yearly rhythm of horses with roughly half a year at work in the forest and living inside and the other half free in the pastures was based on practicality. It was an adaptation to living in the forests of the North. The close relationship came about from humans sharing a living space with horses. During the cold winters, horses brought warmth, companionship, and safety to people. Animals were an integral part of the household, aiding in people's very survival. Birth, life, and death were shared within the hybrid human–animal household. Hence, the northern location and cold weather in combination with folk knowledge resulted in an intimate and caring human–horse relationship.

In the context of nineteenth-century human–horse cohabitation, horses were active agents. The agency of the horse can be described as action and communication with humans and other animals. These agents co-constructed their world together. The lifestyle of the swidden cultivators made it possible to give horses a large amount of freedom and responsibility. Horses were able to act on their intentions with much more freedom than is possible in today's society. Today, horsekeeping of such a nature could even be considered animal abuse. Horses may be better looked after today, but at the same time there is something missing, because the possibilities of a close relationship with animals are constrained by haste and the rapid pace of life. However, a need to be with animals can be seen today in increasing numbers of pets and horses and in the popularity of animal-assisted therapy.

The conception of horses held by the nineteenth-century swidden cultivators is very different from today's views. It was influenced by a nature-bound subsistence and worldview and by everyday practices of keeping horses. Even though horses are sometimes allowed inside houses in Finland today to receive treats, and Finns speak warmly about these visits and consider them a sign of trust between the family and their horse, the idea of human–horse cohabitation is an alien one today. It is a good example of how much a society can change in not much more than a century. It is also an excellent reminder that studying past societies and their practices offers us a glimpse of the variety of human–animal relationships.

Notes

1 See, for example, Anderson and Nuttall, *Cultivating Arctic Landscapes*; Heikkinen, Sarkki, and Nuttall, "Users or Producers of Ecosystem Services?"; Cassidy, "Lives with Others."
2 See, for example, Kaarlenkaski, "Of Cows and Women."
3 Vilkuna, *Suomalaisen karjasuojan vaiheita*; Heikinmäki, "Arkaistisia piirteitä suomalaisessa asumisessa."
4 Armstrong Oma, "Between Trust and Domination."
5 Istomin and Dwyer, "Dynamic Mutual Adaptation."
6 Morris, *The Power of Animals*, 2; Leinonen, *Palvelijasta terapeutiksi*, 29.
7 Theodossopoulos, "Care, Order and Usefulness," 15–6.
8 Armstrong Oma, "Human-Animal Meeting Points," 163.
9 For most of the nineteenth century and beyond (from 1809 until gaining independence in 1917), Finland was an autonomous state of the Russian Empire, called the Grand Duchy of Finland. Before that, it was a part of Sweden for centuries. See Solsten and Meditz, *Finland.*
10 Sarmela, *Finnish Folklore Atlas*, 34.
11 Gummerus and Alfthan, *Lyhyt hevoshoito-oppi*, 41.

12 Topelius, *Maamme kirja*, 100, 102.
13 "Suomen kansan vanhat runot" (SKVR) is a collection encompassing folklore from 1564 onwards.
14 Vilkuna, *Suomalaisen karjasuojan vaiheita*; Heikinmäki, "Arkaistisia piirteitä suomalaisessa asumisessa."
15 Kaarlenkaski, "Cattle Tending," 28.
16 Ingold, *The Perception of the Environment*.
17 Myllyntaus, "Summer Frost," 96.
18 Sarmela, *Finnish Folklore Atlas*, 424, 434, 440.
19 SKS KRA. E253 Oulasmaa, 1960. The code E253 refers to collection E (of ethnographic material), Oulasmaa is the surname of the collector, and 1960 is the year in which the ethnographic material was collected.
20 Lehikoinen, *Ole siviä sikanen*, 331.
21 Ibid., 78.
22 Haavikko, *Hevonen taiteessa*, 108.
23 Anna huolia hevosen/Hevosell on luu lujempi,/Luu lujempi, pää kovempi. SKVR VII2. 1744 Suojärvi. Europaeus, 1846. English translation by Riitta-Marja Leinonen.
24 Vilkuna, *Suomalaisen karjasuojan vaiheita*, 15; Heikinmäki, "Arkaistisia piirteitä suomalaisessa asumisessa," 78–9.
25 Vilkuna, *Suomalaisen karjasuojan vaiheita*, 52.
26 Ibid., 12–6; Heikinmäki, "Arkaistisia piirteitä suomalaisessa asumisessa," 85–7.
27 Armstrong Oma, "Between Trust and Domination," 182.
28 SKS KRA. E42 Brandt, 1893; SKS KRA. E Paulaharju 27497, 1916; SKS KRA. E Paulaharju 27500 and 27501, 1928.
29 Alfthan, *Suomalainen hevonen*, 19.
30 SKS KRA. E Paulaharju 27497, 1916.
31 SKS KRA. E 177 Korosuo, 1947.
32 Vilkuna, *Suomalaisen karjasuojan vaiheita*, 15; SKS KRA. E42 Brandt, 1893; SKS KRA. E Paulaharju 27497, 1916; SKS KRA. E Paulaharju 27500 and 27501, 1928.
33 Leinonen, *Palvelijasta terapeutiksi*, 141, 145.
34 SKS KRA. E42 Brandt, 1893; SKS KRA. E Paulaharju 27497, 1916; SKS KRA. E Paulaharju 27500 and 27501, 1928.
35 Vilkuna, *Suomalaisen hevosen historia*, 75; SKS KRA. E Paulaharju 27497, 1916; SKS KRA. E Korosuo, 1947.
36 Vilkuna, *Suomalaisen karjasuojan vaiheita*, 13.
37 YLE Areena Radio, "Suomenhevonen."
38 Nieminen, *Sata vuotta eläinten puolesta*, 45.
39 Gummerus and Alfthan, *Lyhyt hevoshoito-oppi*, 82.
40 Armstrong Oma, "Between Trust and Domination," 177.
41 McFarland and Hediger, "Approaching the Agency," 3, 11; Wathan, "Social Communication."
42 Wathan, "Social Communication;" Notzke, "An Exploration into Political Ecology," 393; Wolch, "Anima Urbis," 726; Bekoff, "Animal Passions and Beastly Virtues," 71–104.
43 Notzke, "An Exploration into Political Ecology," 393.
44 Shaw, "The Torturer's Horse," 146–67.
45 Weil, *Thinking Animals*, 58–60.
46 Wells, *Pandora's Seed*.
47 Wathan, "Social Communication," 6.
48 Maurstad, Davis, and Cowles, 'Co-Being and Intra-Action," 322–35; Brandt, "A Language of Their Own," 299–316.
49 McDonnell, *A Practical Guide to Horse Behavior*, 72.
50 Armstrong Oma, "Human-Animal Meeting Points," 172.

51 Armstrong Oma, "Between Trust and Domination," 177.
52 Weil, *Thinking Animals*, 59.
53 Armstrong Oma, "Between Trust and Domination," 177.
54 Leinonen, *Palvelijasta terapeutiksi*, 37.
55 SKS KRA. E122 Kallio, 1936; Vilkuna, *Suomalaisen karjasuojan vaiheita*.
56 Löfgren, "The Nature Lovers," 84–5.
57 Vilkuna, *Suomalaisen karjasuojan vaiheita*, 14–5.
58 Anttonen and Viljanen, "Mary Douglas ja ajattelun yhteisöllisyys," 15.
59 Löfgren, "The Nature Lovers," 77–9.
60 See the interview given by Marianne Lien in Isaksen, "Where Does Nature End and Culture Begin?"
61 Sarmela, "Meritokratian eläinkuvia," 110.
62 Armstrong Oma, "Human-Animal Meeting Points," 164.

Bibliography

Ethnographic and Folklore Materials

The Finnish Literature Society (SKS)
- The Folklore Archives (KRA), Ethnographic material (E)
- Suomen kansan vanhat runot (SKVR). http://skvr.fi/.

Literature

Alfthan, Axel. *Suomalainen hevonen*. Helsinki: Otava, 1911.

Anderson, David G., and Mark Nuttall, eds. *Cultivating Arctic Landscapes: Knowing and Managing Animals in the Circumpolar North*. New York: Berghahn Books, 2004.

Anttonen, Veikko, and Anna Maria Viljanen. "Mary Douglas ja ajattelun yhteisöllisyys." Introduction to *Puhtaus ja vaara*, by Mary Douglas, 7–25. Tampere: Vastapaino, 2000.

Armstrong Oma, Kristin. "Between Trust and Domination: Social Contracts between Humans and Animals." *World Archaeology* 42 (2010): 175–87.

Armstrong Oma, Kristin. "Human-Animal Meeting Points: Use of Space in the Household Arena in Past Societies." *Society & Animals* 21 (2013): 162–77.

Bekoff, Marc. "Animal Passions and Beastly Virtues: Cognitive Ethology as the Unifying Science for Understanding the Subjective, Emotional, Empathic, and Moral Lives of Animals." *Zygon* 41 (2006): 71–104.

Brandt, Keri. "A Language of Their Own: An Interactionist Approach to Human-Horse Communication." *Society & Animals* 12 (2004): 299–316.

Cassidy, Rebecca. "Lives with Others: Climate Change and Human–Animal Relations." *Annual Review of Anthropology* 41 (2012): 21–36.

Gummerus, Kaarlo, and Axel Alfthan. *Lyhyt hevoshoito-oppi*. Porvoo: WSOY, 1918.

Haavikko, Ritva. *Hevonen taiteessa, runoudessa, historiassa*. Helsinki: WSOY, 2003.

Heikinmäki, Maija-Liisa. "Arkaistisia piirteitä suomalaisessa asumisessa." In *Kansa kuvastimessa: Etnisyys ja identiteetti*, edited by Teppo Korhonen and Matti Räsänen, 62–91. Helsinki: SKS, 1989.

Heikkinen, Hannu I., Simo Sarkki, and Mark Nuttall. "Users or Producers of Ecosystem Services? A Scenario Exercise for Integrating Conservation and Reindeer Herding in Northeast Finland." *Pastoralism: Research, Policy and Practice* 2 (2012). doi: 10.1186/2041-7136-2-11.

Ingold, Tim. *The Perception of the Environment: Essays in Livelihood, Dwelling and Skill*. London: Routledge, 2000.

Isaksen, Karoline Kvellestad. "Where Does Nature End and Culture Begin?" Centre for Advanced Study at the Norwegian Academy of Science and Letters. Accessed

August 24, 2016. http://cas.oslo.no/full-width-article/where-does-nature-end-and-culture-begin-article1830-1082.html.

Istomin, Kirill V., and Mark J. Dwyer. "Dynamic Mutual Adaptation: Human–Animal Interaction in Reindeer Herding Pastoralism." *Human Ecology* 38 (2010): 613–23.

Kaarlenkaski, Taija. "Of Cows and Women: Gendered Human-Animal Relationships in Finnish Agriculture." *Relations: Beyond Anthropocentrism* 2 (2014): 9–26.

Kaarlenkaski, Taija. "Cattle Tending in the 'Good Old Times': Human–Cow Relationships in Late Nineteenth-Century and Early Twentieth-Century Finland." In *Affect, Space and Animals*, edited by Jopi Nyman and Nora Schuurman, 25–39. London: Routledge, 2016.

Lehikoinen, Heikki. *Ole siviä sikanen. Suomalaiset eläinuskomukset*. Helsinki: Teos, 2009.

Leinonen, Riitta-Marja. *Palvelijasta terapeutiksi: ihmisen ja hevosen suhteen muuttuvat kulttuuriset mallit Suomessa*. Oulu, Finland: University of Oulu, 2013.

Löfgren, Orvar. "The Nature Lovers." In *Culture Builders: A Historical Anthropology of Middle-Class Life*, edited by Jan Fryckman and Orvar Löfgren, 42–87. New Brunswick, NJ: Rutgers University Press, 1987.

Maurstad, Anita, Dona Davis, and Sara Cowles. "Co-Being and Intra-Action in Horse-Human Relationships: A Multi-Species Ethnography of Be(com)ing Human and Be(com)ing Horse." *Social Anthropology* 21 (2013): 322–35.

McDonnell, Sue. *A Practical Guide to Horse Behavior: The Equine Ethogram*. Lanham, MD: Blood-Horse Publications, 2003.

McFarland, Sue, and Ryan Hediger. "Approaching the Agency of Other Animals: An Introduction." In *Animals and Agency: An Interdisciplinary Exploration*, edited by Sue McFarland and Ryan Hediger, 1–20. Boston: Brill, 2009.

Morris, Brian. *The Power of Animals: An Ethnography*. Oxford: Berg, 2000.

Myllyntaus, Timo. "Summer Frost: A Natural Hazard with Fatal Consequences in Preindustrial Finland." In *Natural Disasters, Cultural Responses: Case Studies toward a Global Environmental History*, edited by Christof Mauch and Christian Pfister, 77–102. Plymouth: Lexington Books, 2009.

Nieminen, Hannu. *Sata vuotta eläinten puolesta. Kertomus Suomen Eläinsuojeluyhdistyksen toiminnasta 1901–2001*. Helsinki: Suomen Eläinsuojeluyhdistys, 2001.

Notzke, Claudia. "An Exploration into Political Ecology and Nonhuman Agency: The Case of the Wild Horse in Western Canada." *The Canadian Geographer/Le Géographe canadien* 57 (2013): 389–412.

Sarmela, Matti. "Meritokratian eläinkuvia." In *Eläin ihmisen mielenmaisemassa*, edited by Henni Ilomäki and Outi Lauhakangas, 174–92. Helsinki: SKS, 2002.

Sarmela, Matti. *Finnish Folklore Atlas: Ethnic Culture of Finland 2*, translated by Annira Silver. Helsinki, 2009. Accessed August 25, 2016. www.kotikone.fi/matti.sarmela/folkloreatlas.pdf.

Shaw, David Gary. "The Torturer's Horse: Agency and Animals in History." *History and Theory* 52 (2013): 146–67.

Solsten, Eric, and Sandra W. Meditz, eds. *Finland: A Country Study*. Washington, DC: GPO for the Library of Congress, 1988. Accessed August 13, 2015. http://countrystudies.us/finland/.

Theodossopoulos, Dimitros. "Care, Order and Usefulness: The Context of the Human-Animal Relationship in a Greek Island Community." In *Animals in Person: Cultural Perspectives on Human-Animal Intimacies*, edited by John Knight, 15–36. Oxford: Berg, 2005.

Topelius, Zachris. *Maamme kirja*. Porvoo: WSOY, (1875, 1899) 1983.

Vilkuna, Kustaa. *Suomalaisen hevosen historia*. Offprint, Suomen Museo, 1958.

Vilkuna, Asko. *Suomalaisen karjasuojan vaiheita*. Helsinki: Suomen muinaismuistoyhdistys, 1960.

Wathan, Jennifer. "Social Communication in Domestic Horses: The Production and Perception of Facial Expressions." PhD diss., University of Sussex, 2015.

Weil, Kari. *Thinking Animals: Why Animal Studies Now?* New York: Columbia University Press, 2012.
Wells, Spencer. *Pandora's Seed: The Unforeseen Cost of Civilization*. New York: Random House, 2010.
Wolch, J. "Anima Urbis." *Progress in Human Geography* 26 (2002): 721–42.
YLE Areena Radio. "*Suomenhevonen*." Accessed December 4, 2012. http://areena.yle.fi/radio/1754933.

5 Spectacles of Modern Companionship

Men, Dogs and Early Finnish Dog Shows

Taina Syrjämaa

Introduction

Hurra, Rolff, Diana Pil, Dulcimer, Kairi, Rehvi and some 150 other dogs were the center of attention in May 1891 when the first national dog show took place in Finland. The senior canine was 12-year-old Hurra, a Swedish-born crossbreed hound, who at the time was living in the household of the trader M.E. Fazer in Helsinki. Rolff and Diana Pil were among the youngest dogs on show. The one-year-old Småland-Finnish[1] Hound siblings lived in the household of an engineer in Naantali, a small town in southwest Finland.[2]

The three-day exhibition brought together canine individuals with different backgrounds. Most lived in southern Finland, not too far from Helsinki, but many originated from other districts and countries. For example, Dulcimer, a three-year-old Foxhound owned by Daniel Wadén, the president of the dog show committee, had English pedigree and had arrived in Finland two years earlier from a large British kennel.[3] Wadén also owned the red-gray Kairi, a Finnish Barking Bird Dog, who came from a peasant household in rural northern Finland. Another representative of the same breed – later to be known as Finnish Spitz – was Rehvi, who originated from an even more peripheral district, the subarctic Russian Kola Peninsula. He had been acquired in his remote home region by a Karelian trader, who transported him to a Finnish manor house where he resided with his new human companion, Walter von Konow, a university student.[4]

Nineteenth-century urbanization and modernization brought many changes to the lives of humans and animals, as well as to their mutual relations. Urbanization rendered unnecessary and even impossible many practices of co-living, which had been self-evident in rural societies, where the daily toils of the majority of people were directly connected to livestock and where the wilderness was in the immediate vicinity. Contemporaneously, the utilization of animals acquired industrial dimensions. For example, the use of draft animals, which powered the growth of traffic and industry, grew massively. Modern science also widely practiced vivisection on animals.[5] At the same time, some other animals were privileged. Harriet Ritvo has shown how the booming trend for pets evolved in parallel with the growing conviction that humans had mastered nature. When people – especially those who were wealthy – had confidence in the human ability to overcome and command nature, they were able to regard animals with tenderness and affection.[6] In many aspects these were transnational transformations, yet there were also national and regional differences.

This chapter explores a form of modern companionship of dogs and humans – in practice, men – and their shared history as manifested and forged in dog shows, especially in regard to the very first nationwide Finnish dog show. I examine the mutuality and relational agency by investigating, firstly, which dogs and men were most directly involved, and, secondly, how dogs were simultaneously specimens of breeds and individuals with divergent backgrounds and experiences. Thus, the chapter brings individuals to the fore, whether they be human or canine.

The dog show was a transnational phenomenon and formed part of a most powerful mass medium of its time: exhibitions. Dog shows became popular in Great Britain in the mid-nineteenth century, in the wake of all manner of expositions, and they soon began to take place across Europe and North America.[7] They constituted part of a widespread culture of display that was fundamentally connected to the production and circulation of knowledge and which celebrated and supposedly accelerated the much cherished notion of progress. As social events, exhibitions, including dog shows, manifestly belonged to a modern, urban and bourgeois way of life.[8]

From the very beginning, the Finnish Kennel Club, tailored according to the British model and established in 1889, aimed at organizing dog shows in order to promote breeding.[9] The first nationwide show in 1891 was on a grand scale when compared to the small, rather random sections previously devoted to dogs in agricultural fairs.[10] However, from an international perspective, the Helsinki dog show was tiny,[11] but this does not seem to have spoiled the excitement for Finnish enthusiasts.

In a circular, the Finnish Kennel Club emphasized how arranging a dog show was a practice widely adopted in civilized world.[12] Thus, it was suggested that a dog show was a necessity for Finland if the country aspired to be recognized as a modern and cultured land. Showing national competence was a topical issue in late nineteenth-century Finland, due to intensified nationalistic aspirations, but other reasons were also behind the popularity of dog shows. In this context, the appreciation of dogs as a species was crucial. This attitude was not self-evident. A decade earlier, when seventeen dogs had been displayed in a national agricultural fair, a newspaper commentator had explicitly called the section worthless, as dogs hardly served any purpose in "filling a crib of rye."[13]

The examination in this chapter is located in both the transnational context of the exhibition medium and dog breeding, as well as in the specific northern context of Finland. Finland was among those countries that industrialized relatively late, even though the pace of industrialization began to accelerate in the 1870s in key fields, such as in timber-processing industries. However, for a long time Finland remained a country of small farmers with a relatively slow rate of urbanization.[14] The entire country, and especially the central and northern hinterlands, remained much more sparsely populated than most European countries.

Sharing Agency and History

The empirical analysis in this chapter is based on a close reading of dog show catalogs, other material compiled by the Finnish Kennel Club as well as media coverage of the dog shows. Dorothee Brantz has noted that "writing the history of animals demands negotiating our desire to recover the historical lives of animals vis-à-vis the fact that all of the available records of those lives have been produced by humans."[15] Because of human bias, historical sources more readily highlight human agency and downplay or ignore that of animals. For example, the show catalogs may initially only seem to provide evidence of a human attempt to organize, govern and manipulate dog breeds.

Éric Baratay suggests that the most decisive factor is to recognize and admit that humans and animals always live together by necessity: sometimes more willingly, sometimes by force, but that they always influence each other. Once this idea and the consequent widening of the scope of history are approved, the more complex reality including non-human animals eventually becomes "easy to find, to observe, to write."[16] In a similar vein Susan Nance has remarked that a key issue is to see all history as interspecific.[17]

When dog show catalogs are analyzed from this perspective, a new panorama appears, which includes interacting agents from different species. The catalog entries typically consist of basic information concerning a dog and a human, indicating their names, the dog's age, breed and pedigree as well as the human's profession and place of residence. Sometimes the catalogs also included information on how and when the dog had been acquired, or regarding the circumstances in which the dog had lived prior to being purchased by its current owner. Thus, these documents actually provide – albeit in a fragmentary and patchy manner – biographies of the dogs. Furthermore, they emphasized the close connection between human and canine individuals.

This approach highlights the intertwinement of human and canine lives, thereby reducing the seemingly fundamental differences and blinding degree of otherness. This, in turn, makes the dogs visible as sentient, living individuals instead of simply being objects of human action and imagination. It also encourages us to examine prosaic narrations, such as overtly anthropocentric media reports, so that analysis does not end with human impressions, but extends to human–animal interactions and to some extent to animal experience.

The capacity to act varies and the co-living of humans and dogs is fundamentally asymmetric and unequal. Also in the dog show context, disparities are evident: owner and owned, exhibitor and exhibit(ed). In his classic and influential book *Dominance and Affection*, Yi-Fu Tuan has depicted the relationship of humans and tame animals by emphasizing the unavoidable supremacy and cruelty of humans, as well as animals' submission.[18] Tuan discloses important – and disturbing – aspects of human–animal relations, but his view does not allow much, if any, attention to animal agency. More recently, non-human agency in general has attracted much consideration and the previously automatically assumed focus on humans has been seriously questioned.[19] Similarly in human–animal studies the focus has started to shift towards exploring mutuality: how human and non-human animals both matter and both influence each other, thus creating shared lives.[20] As Susan Nance has compellingly shown in her study on nineteenth-century circus elephants, animals live in a kind of parallel reality in which human social institutions and structures are not meaningful to them as such. However, this framework fundamentally influences their lives and in turn the animals affect these institutions and people.[21] The starting point in this chapter is that agency – whether animal or not – is essentially relational and no actor is immune to the intended or unintended influence of others.

Protagonists in Helsinki

In May 1891, a riding hall, the Guard Manège, became the temporary venue for approximately 150 dogs owned by 98 persons, as well as entries belonging to a hunting club, a military unit and a dog trading company.[22] As in any serious exposition, the exhibits – the dogs – were classified and consequently displayed in a seemingly good order.[23] While many of the 20 classes of the show had only one or a few participants, there were also quite large groups, such as 16 Finnish Barking Bird Dogs, 13 Pointers and 12 German Bird Dogs. The largest group by far were 57 cross-breed hounds. A closer examination of the three most eager human participants and their 23 displayed dogs allows us to get an idea of the inner circle of the event.

Two of these enthusiasts were noblemen, Baron Ludvig Standertskjöld and Captain Hjalmar Rosenbröijer, and the third was Daniel Wadén, the president of the organizing committee who, by profession, was a pioneering figure in the Finnish telephone business. All possessed dogs of foreign origin. Rosenbröijer specialized in breeding "Kostroma-Polish"[24] hounds and seems to have kept some sort of kennel. He presented Saschiga and Ukko, both of whom were five years old, with puppies some of whom had already been sold. He had been in the dog business for quite some time and had exhibited his hounds in St. Petersburg.[25]

Standertskjöld and Wadén, on the other hand, spent considerable funds on buying pedigrees. They were interested in several breeds, including Dachshunds, Fox Terriers, Setters and St. Bernards. Indeed, Standertskjöld presented six young "imported" dogs, including Caesar, a Great Dane, who had been acquired in a German kennel. He had also purchased Diana, a St. Bernard, from St. Petersburg and two Icelandic Herding Dogs, Pajas and Blanka, from northern Sweden. Furthermore, his Gordon Setters, named Devonshire Duke and Devonshire Countess, were pedigree dogs from England. Many of Wadén's dogs were also of foreign origin: in addition to the above mentioned Dulcimer, his two Pointers, Milton Princess and Milton Totnes, had full English pedigree.[26]

During the nineteenth century dog breeding was a fashionable and fickle enterprise. Many new breeds were introduced, alongside plentiful modifications to existing breeds. Practically all well-known dog breeds were foreign from the

Finnish perspective and, thus, it was natural that the most highly appreciated individuals were of foreign origin. The wish to participate in the transnational trend resulted in a notable degree of canine mobility. Many dogs were imported from Britain, as the country was a forerunner in modern dog breeding.[27] Dog mobility was, however, a more versatile phenomenon, as the above examples show in terms of their varied countries of origin. It is quite reasonable, for example, that Standertskjöld's Caesar was acquired in a German kennel, as the Great Dane was a breed developed in Germany. Large foreign kennels and cosmopolitan centers had many internationally renowned breeds on offer, but in specific cases dogs were sought in remote rural locations, such as Lapland. Thus, canine geography had mainstreams and core arteries, but it was also dotted with a great number of divergent paths.

The instability of the rapidly evolving field of dog breeding and the lack of pedigree dogs with long certified family trees in Finland is reflected in the great number of unofficial breeds. When the breed of a dog was difficult or impossible to label, it was given an ad hoc breed denomination referring to the geographical region of origin. In the Helsinki show, for example, certain dog breeds were referred to as Anglo-Swiss, Anglo-Swedish-Augustenburgian, German-Finnish and Anglo-German-Swedish.[28] These appellations also denote dog mobility, although they do not necessarily refer to the recent transfer of an individual, but to the relocations of several individuals belonging to different generations.

Figure 5.1 An example of dog mobility: the English-born Dulcimer and Gladness depicted after their arrival in Finland. *Sporten*, April 15, 1889.

The foreignness of the dogs on show was usually highly appreciated. The arrival of a fine – and presumably expensive – individual could even be newsworthy: when the English-born Dulcimer arrived in Finland with a female dog, Gladness, the pair adorned the front cover of the country's sole sporting journal.[29] (See Figure 5.1). More generally speaking, the conception of Finnish marginality and backwardness was strong and the appraisal of foreignness was quite common in Finland.[30] Hence, imported dogs were also seen as an important link connecting Finland to the most up-to-date centers and as prerequisites for modernity and progress.

However, the lure of the foreign was intertwined with burgeoning nationalism: dog breeds could be perceived as being the epitome of a nation. Finnish efforts to develop a national dog breed that met international standards focused on two kinds of hunting dog. Daniel Wadén promoted one of them, namely the Finnish Barking Bird Dog, or Finnish Spitz, by exhibiting Kairi in a number of shows.[31] She competed in a class in which dogs were not imported from fine foreign kennels, but were brought from rural parts of northern Finland, Swedish Lapland or northwest Russia and who had absolutely no known pedigree history. Neither the parents nor the previous owners of these dogs were known or were considered worthy of documentation, but the dogs themselves were seen as fine representatives of an authentic Finnish breed. The supposed Finnishness of these dogs was celebrated, despite some dogs being imported from neighboring countries, as was the above-mentioned Rehvi.

What about the humans who organized and attended the show? What kind of people were the "hunters and friends of dogs" to whom the show was aimed?[32] Standertskjöld, Rosenbröijer and Wadén were of high, yet divergent, social standing, as was the case with most of those who enrolled their dogs in the show. The long list of the titles and professions of the dog owners includes eight barons, seven military professionals and a variety of high public servants. However, the most numerous single group were merchants, with at least ten individuals stating this as their profession. There were also five doctors, three engineers, two professors, two pastors, a rector, a bank director, a pharmacist, a veterinarian and university students. Those who owned pedigree dogs and were interested in the dog show as a forum were evidently well-off citizens. Yet, they still formed a rather heterogeneous group. Simultaneously, many kinds of potential "friends of dogs" were missing from the event, including women and peasants.

Only three dogs had female owners: Tello, a Newfoundland Hound, Rappo, an Irish Terrier and Trogen who was first classified as a Finnish Spitz and later as a Lapponian Herder. Female participation also remained marginal in the subsequent shows, but those few who did participate could be quite successful. Miss Nybergh's Trogen and widow Fröberg's Tello were awarded prizes and were also entered in consecutive shows.[33] It is not clear how common it was for women to own dogs, but it is evident that many women co-lived with dogs.[34] However, this kind of co-living remained rather invisible in the context of dog shows and kennel clubs, in which the emphasis was on hunting dogs.

Hunting as a pastime had gained popularity in Finland in the decades leading up to the first dog show. A national hunting club had been founded in the 1860s and a great number of regional ones were subsequently established. In Finland, hunting had traditionally formed an important part of a peasant's livelihood and it was still common for the rural population to hunt. However, new legislation enacted unprecedented hunting regulations, such as closed seasons for various species of game. These laws were based on a conception of hunting as a sport rather than as a rural livelihood. The peasant manner of hunting was becoming marginalized and in some instances even classified as illegal poaching.[35] In the same manner as their hunting practices were being disregarded, they and their dogs were mostly excluded from shows.

As in the cases of Kairi and Rehvi, some dogs living in rural areas were considered to possess innate qualities that could assist in promoting Finnish breeds. Consequently, such dogs were acquired by wealthy urbanites or manor house owners in southern Finland. Similar canine individuals, who were still owned by peasants, were also sought after for the show. An enrollment discount was arranged for peasants' dogs in the show, as well as for their rail transportation.[36] No discount was offered to their human companions, whose presence was evidently not anticipated. In fact, in some instances, dogs could be more directly involved in a modern, urban practice than their owners.

At the Exhibition

> When one enters, it all makes a very nice impression. Though the ears are disturbed by howling, which is anything but dulcet, one soon forgets to pay attention to these inharmonious sounds and instead views more closely these quadrupeds, which are treated with plenty of patting and caresses by the visitors.[37]

Reporters emphasized the overwhelming soundscape, dominated by incessant and loud barking, when they first entered the dog show arena.[38] In addition to the cacophonous soundscape, some, less prudish, commentators noted that the venue stank. The dogs' presence in the show was, naturally, evident. It was the rationale of the entire event, but it led to interactions between non-human animals and humans that were far from familiar in most other exhibitions or social occasions attended by the urban public. It was also a very unfamiliar setting for most of the dogs.

Tracing the experiences of a single dog remains outside the scope and the possibilities of this chapter, but human descriptions of dogs' behavior suggest that it was rather a stressful event for the canines. Drawing on these descriptions, Helena Telkänranta, a biologist specializing in animal behavior and cognition, has identified a number of different ways in which dogs expressed anxiety. First, barking and howling were the most easily noticeable traits and were the most often mentioned actions. Large- and medium-sized dogs, who dominated the show, tend to be low-voiced, but the descriptions of the show emphasized high-pitched barking, which denotes nervousness and fear. Howling, on the other hand, is a way a dog attempts to rejoin its family members when it has been separated from them. Whining seems to have been a rarer manner of communication for the dogs in the exhibitions. This form of behavior denotes a dog's distress in an uncomfortable situation, but it is used for attracting the attention of a trusted person. When such a person is not nearby, a dog may prefer to howl or bark.[39]

It is no wonder that dogs were anxious during the show. Practically all of them were without their familiar human companions in the exhibition premises. Many were also separated from other canine family members. Some dogs did share a stall with their siblings or mates, but many were fastened alone in a stall in the midst of a myriad of strange odors, sounds and curious human onlookers. The dogs who lived in Helsinki could be taken home for the night, but others remained in their stalls all the time. Some dogs had endured long journeys, like the Finnish Spitz Kyttä, who had arrived from Ii, situated north of Oulu. The railway up to Oulu had only been inaugurated in 1886, and the 752 kilometer journey lasted over 37 hours.[40] It is not known whether Kyttä's owner traveled to Helsinki, but dogs were generally transported as freight in trains, which entailed wearing a muzzle in respective boxes. In short, the journey must have been an ordeal.

Some dogs were reported to have sought human contact, or at least resigned themselves to being petted. Among these was one-year-old, English-born Grogg, a "golden and black striped bulldog."[41] Some other dogs were reported to be very calm and passive, which divided the commentators' views. Some interpreted this as a sign of good education, while some considered the dogs as looking sad and weary.[42] Telkänranta notes that dogs' conduct in stressful situations varies and some individuals may be passive and withdrawn. In some cases such behavior possibly hinted at previous exhibition experiences, which may have partly mitigated a dog's fear as the circumstances would have been less unfamiliar. Yet, the show would still remain a repulsive and stressful event.[43]

Dog shows were highly exceptional events for almost every dog, but the degrees of unfamiliarity varied according to a dog's background and usual living circumstances. Andria Pooley-Ebert has interestingly shown how experiences and behavior varied between horses living in rural Illinois and in Chicago at the turn of the twentieth century. She connects the differences in horses' modes of action from resistance to compliance to their respective living environments, which favored divergent modes of human–animal co-living.[44] The available sources do not here allow a detailed comparison between dogs of different backgrounds, but it is evident that some dogs had, for example, experience with living with numerous other canines. At the other extreme were the dogs who had lived as pets in households. Furthermore, the city as an environment, with its clamor and odors, was obviously less familiar to some of the dogs. It seems likely, for example, that Kyttä only knew countryside villages, peasant farms, and forests prior to being in the Helsinki dog show. To him and those like him the dog show experience consisting of a railway journey, the noise of a military band, the taste of industrially produced dog biscuits and the presence of numerous other dogs and unknown humans was extremely unfamiliar and potentially stressful.

Despite some comments on gloomy looking dogs, contemporaries did not question the conditions of the dog show. The riding hall offered a spacious area for the audience to circulate and to get close enough to the dogs that they could study and even touch and cuddle them. Visitors often commented on a dog's physical features, which were important when seeking to define and develop breeds. The dogs were also supposed to display certain traits linked to the temperament and mental capacities of their breed. Accordingly, a reporter noted that Standertskjöld's Icelandic Herding Dogs, Pajas and Blanka, were "smallish, completely white in color, thick in coat, with pointed muzzles" and that they "seem to have a lot of wit and a friendly and frisky character."[45] In general, visitors sought to interact with the dogs and considered themselves capable of judging a dog's intelligence and nature.

Favorable looks, however, did not necessarily mean that dogs had any discernible talent for hunting, which required an ability to discover and chase game, and to communicate with the hunter and potentially with other hunting dogs. Eventually, some people argued that a different kind of setting was necessary: "[a dog] should be seen in the company of his master in the woods or meadows, where he gallops around in a lively manner, and has an eager desire to hunt."[46] To compensate for the shortcomings of the dog show environment, separate field trials were planned for soon after the exhibition. However, due to the remote site and format of the event, these trials were not accessible to the general viewing public and were limited to a highly exclusive number of dogs and dog owners.

Dog fanciers flaunted any prizes their dogs won in shows and trials in a similar manner to companies that received awards for their products. A prosaic dimension of the exhibition was that a show prize could raise a dog's sale value. Yet, only a few exhibitors actually sold dogs.[47] Hence, they must have perceived that exhibiting a dog would secure other benefits. In his highly critical assessment of American society and modern consumption in the late 1890s, Thorstein Veblen classified domesticated animals, such as dogs and fast horses, as potential "items of conspicuous consumption." He argued that dogs were well suited to be in this category because they served no useful industrial purpose and, secondly, because specific kinds of dog were relatively scarce and their prices were high due to constant changes in breeding trends.[48] Although the Helsinki exhibitors were far removed from the opulence of the Americans which were the focus of Veblen's criticism, certain kinds of dogs and practices were apparently socially appreciated.

Yet, interspecies relations are much more complex than merely being based on more or less conscious human calculations of utility. Even the disapproving Veblen did not deny that humans and dogs could potentially develop deeply affectionate attachments. According to him, people tend to equate pecuniary value to other positive qualities, such as beauty. Thus, an expensive dog was a beautiful dog, although many of these trendy products of breeding could have also been seen as deformed creatures. However, as it was socially respectable to have a pet, it was possible that an affectionate tie would eventually be formed.[49]

The Finnish reading public could get an inkling of extravagant dog stories from overseas as newspapers referred to dogs who drank from silver plates.[50] One of the dogs displayed in Helsinki, a Scottish Herding Dog, was actually said to be the offspring of a dog bought by an American at an astronomical price.[51] The dog in question was probably Blossom, a three-year-old Collie, who belonged to estate owner Nikolai Heimburger in the vicinity of Helsinki. Such spheres, however, were somewhat detached from local realities, although pride in a dog did play a role in Finnish circles, too.

To enroll a dog in an exhibition already implied that its owner believed that s/he was a fine dog. Some sold, bought and exchanged dogs with seeming ease, but owners and dogs also formed life-long relations. Although most of the dogs who were exhibited were quite young, some of the more senior dogs provide evidence of long-standing comradeship. Dog shows held in public arenas did not make it easy to observe emotional attachment, but for example Walter von Konow, the owner of Rehvi, wrote about him in a warm and benevolent manner. The story not only shows how the Kola-born Rehvi had accepted von Konow's manor house as his home, which he eagerly and loudly guarded, but it also reveals the family's attachment to the dog, who is presented as a founder and leader of his own canine family.[52] A dog could be acquired and exhibited as a good specimen of its breed, but co-living highlighted individuality. Rehvi was rewarded with first prize in 1891, which established his value as an excellent Finnish Barking Bird Dog, as well as his offspring. Yet, Rehvi's breed formed just one dimension. Contemporaneously, it was all about an individual with his habits and experiences, some of which derived from a life spent in a subarctic village. Others were gained when Rehvi co-lived and hunted with the von Konows and yet others while travelling and being part of the chorus of barking dogs in the Guard Manège in May 1891.

Conclusion

Arranging and participating in exhibitions was a well-established practice, which was considered to be beneficial for society at large and for citizens who aspired to educate and improve themselves. The president of the dog show committee, Daniel Wadén, also had personal experience with the exhibition medium as a businessman.[53] As common as expositions and displays were, there was something quite special about dog shows: they placed living animals at the very center of an urban exhibition and displayed them as companions.

Human actors may seem to be overwhelmingly powerful in the dog show context, as they acquired, owned and bred dogs. Furthermore, they also enrolled them in the exhibition and fastened them into the stalls. The relations of humans and tame animals were – and continue to be – highly unequal. Yet, it is obvious that without dogs there would have been no dog show and they were not anonymous objects. Instead, sentient – and loud – creatures made their own demands and expressed their own preferences. They were there to represent breeds, but contemporaneously they were furry individuals with diverging backgrounds and experiences. Humans reacted to the dogs' participation and presence at the show, and together they constructed an event that was at the pinnacle of dynamically changing practices of interspecies interaction.

Distance was taken from such modes of co-living that were conceived to be old-fashioned or that were attached to the lower estates. These could be well embedded in Finnish natural and social circumstances, but the transnational process of modernization promoted new manners and new values. Southern and coastal regions, which fared relatively well in agriculture, commerce and industry, were particularly involved in international trends. Pedigree dogs, amateur hunters and dog shows long remained outside the culture of northern Finland, which was dictated by the harsh conditions of the natural wilderness.

The dog shows presented dogs – as a species, as breeds and as individuals – that the most well-to-do citizens and the emergent urban middle class wanted to live with. The shows promoted a close modern human–animal relationship at a time when modes and environments were quickly changing. Pedigree dogs became an essential part in a network that sustained nature-based sports, sociability in associations and competing in shows and trials. Furthermore, many pedigree dogs were adopted in families with no direct link to hunting. Thus, they became canine members – and markers – of civilized homes, in which they had no function other than as companions. This opened new opportunities for dogs, and to a growing extent to pets of other species, as well as for humans whose homes and lives they transformed and enriched.

Acknowledgemtent

This research was supported by a grant from the Academy of Finland (274573). I wish to thank Dr. Robert Collis for correcting the language.

Notes

1 Småland is a region in southern Sweden.
2 *Katalog öfver hundar*.
3 "'Dulciner' och 'Gladness,'" *Sporten*, April 15, 1889.
4 [von Konow], "Finska skällande fågelhunden."
5 Ritvo, *The Animal Estate*, 157–65; Kete, "Introduction," 4–6, 18–19; Baratay, *Le point de vue animal*, 107–34. On preindustrial, rural human–animal relations see Donovan, "Provincial Life with Animals," 17–33.
6 Ritvo, *The Animal Estate*, 2–3, 82–121; Ritvo, "The Emergence of Modern Pet-keeping." Also see Kete, *The Beast in the Boudoir*; Brantz, "The Domestication of Empire," 76–9; Baratay, *Le point de vue animal*, 278–90.
7 Ritvo, *The Animal Estate*. 97–101; Grier, *Pets in America*, 29–30.
8 In Finnish and Swedish dog shows were called "dog exhibitions": respectively *koiranäyttely* and *hundutställning*. On the history of the exhibition medium see, for example, Greenhalgh, *Ephemeral Vistas*; Syrjämaa, *Edistyksen luvattu maailma*; Filipová, *Cultures of International Exhibitions*.
9 "Koiranäyttely," *Uusi Suometar*, May 9, 1889; "Kennelklubb och hundutställning," *Sporten*, May 15, 1889; Mäkelä-Alitalo, *Pennusta pitäen*; Unhola, *Elämää koiran kanssa*, 22–6.
10 *Kahdeksas yleinen*; *Yhdeksännen yleisen*.

11 In Britain, dog shows with over 1000 dogs took place on a regular basis. Ritvo, *The Animal Estate* 1987, 97.
12 "Koiranäyttely Helsingissä," *Uusi Suometar*, January 20, 1891.
13 "Kahdeksas yleinen maanviljelysnäyttely Turussa," *Keski-Suomi*, September 10, 1881. On negative and positive attitudes towards dogs, see also Mäkelä-Alitalo, *Pennusta pitäen*, 15–6.
14 Kuisma, *Suomen poliittinen taloushistoria*, 125–38; Kirby, *A Concise History of Finland*, 110–3.
15 Brantz, "Introduction," 5.
16 Baratay, *Le pont du vue animal*, 43–56, quotation 43.
17 Nance, "Introduction," 5–6.
18 Tuan, *Dominance and Affection*, 88–114.
19 Latour, *Reassembling the Social*.
20 The outstanding example in the contemporary context is Haraway, *When Species Meet*. For a historical approach, see Brantz, *Beastly Natures*. Also see McFarland and Hediger, *Animals and Agency*.
21 Nance, *Entertaining Elephants*.
22 *Katalog öfver hundar*.
23 On nineteenth-century quests and principles of ordering and classification, see Ritvo, *The Platypus and the Mermaid*.
24 Kostroma is a Russian city and province northeast of Moscow.
25 Hj. Rbr [Hjalmar Rosenbröijer], "Hundutställningen i S:t Petersburg," *Sporten*, June 15, 1883.
26 *Katalog öfver hundar*, 4, 6–11, 18–19.
27 Ritvo, *The Animal Estate*, 93–7, 102–15; Ritvo, "The Emergence of Modern Pet-keeping," 103–5. Cf. Grier, *Pets in America*, 28–31.
28 *Katalog öfver hundar*, 12–7.
29 "'Dulciner' och 'Gladness,'" *Sporten*, April 15, 1889. These dogs were imported to Finland by Volmari Kanninen, a founding member of the Finnish Kennel Club and its long-time secretary. He documented their respective family trees, both covering three previous generations, in his notebook. National Archive of Finland, Archive of Volmari Kanninen, Kanninen's notebook 1889–1890.
30 Syrjämaa, "Making Difference, Seeking Sameness," 27–40.
31 *Katalog öfver hundar* 12; *Katalog. Finska Kennelklubbens 2:dra*, 19; *Katalog öfver Finska Kennelklubbens 4:de*, 3; Volmari Kanninen, "Finska Kennelklubbens IV allmänna hundutställning," *Finska Kennelklubbens Tidskrift*, September 1896, 16.
32 The circular announcing the first national Finnish dog show emphasized the utility of the event to "all hunters and friends of dogs." The circular was published, for example, in *Finland*, January 20, 1891.
33 *Katalog öfver hundar* 12, 19–20; *Katalog. Finska Kennelklubbens 2:dra*, 47–9; *Katalog öfver Finska Kennelklubbens 4:de*, 55.
34 See, for example, the portraits of women and dogs taken by Augusta Olsson. Uusikaupunki Museum, Augusta Olsson's photograph collection.
35 Lehikoinen, *Tuo hiisi hirviäsi*, 184–91.
36 "Koiranäyttely Helsingissä," *Uusi Suometar*, January 1, 1891.
37 "Kennelklubbens hundutställning," *Finland*, May 23, 1891.
38 "Koiranäyttely," *Päivälehti*, May 24, 1891; "Hundutställningen i gardesmanégen," *Nya Pressen*, May 24, 1891; "Hundutställningen," *Hufvudstadsbladet*, May 25, 1891.
39 Telkänranta's email to the author, August 10, 2015. I wish to thank Helena Telkänranta (University of Helsinki), author of a number of books on animals and animal behavior, for having most kindly read and analyzed those sections of my primary sources that contain descriptions of dogs in shows. Among her publications can be mentioned *Millaista on olla eläin?*

40 Zetterberg, *Yhteisellä matkalla*, 44–7, 80.
41 "Hundutställningen," *Hufvudstadsbladet*, May 25, 1891.
42 For example, "Hundutställningen i gardesmanégen," *Nya Pressen*, May 24, 1891.
43 Telkänranta's email to the author, August 10, 2015.
44 Pooley-Ebert, "Species Agency," 148–65.
45 "Kennelklubin koiranäyttelyyn," *Uusi Suometar*, May 24, 1891.
46 "Hundutställningen," *Hufvudstadsbladet*, May 25, 1891.
47 See, for example, "S:t Bernhards hund's valpar," *Kotka*, October 8, 1891.
48 Veblen, *The Theory of the Leisure Class*, 140–2.
49 Ibid., 142.
50 "Göötepori n näyttelystä," *Päivälehti*, August 26, 1891; "Hundutställningen," *Finland*, August 28, 1891.
51 "Koiranäyttely," *Päivälehti*, May 24, 1891.
52 [von Konow], "Finska skällande fågelhunden."
53 Takkula, "Wadén."

Bibliography

Archival Sources

National Archive of Finland. Archive of Volmari Kanninen. Notebooks.
Uusikaupunki museum, Augusta Olsson's photograph collection.

Newspapers and Magazines

Finland, 1891.
Finska Kennelklubbens tidskrift, 1896.
Hufvudstadsbladet, 1891.
Keski-Suomi, 1881.
Kotka, 1891.
Nya Pressen, 1891.
Päivälehti, 1891.
Sporten, 1883, 1889.
Uusi Suometar, 1889, 1891.

Literature

Baratay, Éric. *Le point de vue animal. Une autre version de l'histoire*. Paris: Seuil, 2012.
Brantz, Dorothee, ed. *Beastly Natures. Animals, Humans, and the Study of History*. Charlottesville, VA and London: University of Virginia Press, 2010.
Brantz, Dorothee, "Introduction." In *Beastly Natures. Animals, Humans, and the Study of History*, edited by Dorothee Brantz, 2–16. Charlottesville, VA and London: University of Virginia Press, 2010.
Brantz, Dorothee. "The Domestication of Empire. Human-Animal Relations at the Intersection of Civilization, Evolution, and Acclimatization in the Nineteenth Century." In *A Cultural History of Animals in the Age of Empire*, edited by Kathleen Kete, 73–93. Oxford and New York: Berg, 2011.
Donovan, Josephine, "Provincial Life with Animals," *Society and Animals* 21 (2013), 17–33.
Filipová, Marta, ed. *Cultures of International Exhibitions 1840–1940*. Farnham, UK: Ashgate, 2015.

Greenhalgh, Paul. *Ephemeral Vistas: The Expositions Universelles, Great Exhibitions and World's Fairs, 1851–1939*. Manchester, UK: Manchester University Press, (1988) 2000.

Grier, Katherine C. *Pets in America: A History*. Chapel Hill, NC: The University of North Carolina Press, 2006.

Haraway, Donna. *When Species Meet*. Minneapolis, MN and London: University of Minnesota Press, 2008.

Kahdeksas yleinen Suomen maanviljelyskokous Turussa 1881. Luettelo. Turku, Finland: G. W. Wilénin ja kumpp:n kirjapaino, 1881.

Katalog öfver hundar, exponerade vid Finska Kennelklubbens Hundutställning i Helsingfors den 23, 24 och 25 maj 1891. Helsingfors, Finland: Finska Kennelklubben, 1891.

Katalog. Finska Kennelklubbens 2:dra Allmänna Hundutställningen i Helsingfors den 13, 14 och 15 maj 1892. Helsingfors, Finland: Finska Kennelklubben, 1892.

Katalog öfver Finska Kennelklubbens 4:de Allmänna Hundutställningen i Helsingfors den 16, 17 och 18 maj 1896. Helsingfors, Finland: Finska Kennelklubben, 1896.

Kete, Kathleen. *The Beast in the Boudoir. Petkeeping in Nineteenth-Century Paris*. Berkeley, CA, Los Angeles and London: University of California Press, 1994.

Kete, Kathleen. "Introduction: Animals and Human Empire," in *A Cultural History of Animals in the Age of Empire*, edited by Kathleen Kete, 1–24. Oxford and New York: Berg, 2011.

Kirby, David. *A Concise History of Finland*. Cambridge, UK: Cambridge University Press, 2012.

Kuisma, Markku. *Suomen poliittinen taloushistoria 1000–2000*. Helsinki: Siltala, 2009.

Latour, Bruno. *Reassembling the Social. An Introduction to Actor-Network-Theory*. Oxford: Oxford University Press, 2005.

Lehikoinen, Heikki. *Tuo hiisi hirviäsi. Metsästyksen kulttuurihistoria Suomessa*. Helsinki: Teos, 2007.

McFarland, Sarah E., and Hediger, Ryan, eds. *Animals and Agency: An Interdisciplinary Exploration*. Leiden, Netherlands and Boston: Brill, 2009.

Mäkelä-Alitalo, Anneli. *Pennusta pitäen. Suomalaisen kennelhistorian ensimmäiset sukupolvet*. Helsinki: SHS, 1998.

Nance, Susan. *Entertaining Elephants. Animal Agency and the Business of the American Circus*. Baltimore, MD: Johns Hopkins University Press, 2013.

Nance, Susan. "Introduction." In *The Historical Animal*, edited by Susan Nance, 1–16. Syracuse, NY: Syracuse University Press, 2015.

Pooley-Ebert, Andria. "Species Agency: A Comparative Study of Horse-Human Relationships in Chicago and Rural Illinois." In *The Historical Animal*, edited by Susan Nance, 148–65. Syracuse, NY: Syracuse University Press, 2015.

Ritvo, Harriet. *The Animal Estate. The English and Other Creatures in the Victorian Age*. Cambridge, MA and London: Harvard University Press, 1987.

Ritvo, Harriet. *The Platypus and the Mermaid, and Other Figments of the Classifying Imagination*. Cambridge MA: Harvard University Press, 1997.

Ritvo, Harriet. "The Emergence of Modern Pet-keeping," In *Social Creatures. A Human and Animal Studies Reader*, edited by Clifton P. Flynn, 96–106. New York: Lantern Books, 2008.

Syrjämaa, Taina. *Edistyksen luvattu maailma. Edistysusko maailmannäyttelyissä 1851–1915*. Helsinki: SKS, 2007.

Syrjämaa, Taina. "Making Difference, Seeking Sameness. Negotiating Finnishness and Foreignness in an Exhibition." In *Nordic Perspectives on Encountering Foreignness*, edited by Anne Folke Henningsen, Leila Koivunen and Taina Syrjämaa, 27–40. Turku, Finland: University of Turku, 2009.

Takkula, Petteri. "Wadén, Daniel Johannes (1850–1930)." In *Kansallisbiografia*. SKS, 2007. Accessed May 2, 2014. www.kansallisbiografia.fi/kb/artikkeli/5235/.

Telkänranta, Helena. *Millaista on olla eläin?* Helsinki: SKS, 2015.

Telkänranta, Helena (biologist specializing in animal behavior and cognition, University of Helsinki), email to the author, 10 August 2015.

Tuan, Yi-Fu. *Dominance and Affection. The Making of Pets*. New Haven and London: Yale University Press, 1984.

Unhola, Kaija. *Elämää koiran kanssa. Suomalainen koiraharrastus ennen ja nyt*. Helsinki: Art House, 2014.

Veblen, Thorstein. *The Theory of the Leisure Class. An Economic Study of Institutions*. New York: The Macmillan Company, (1899) 1912.

[von Konow], W[alter]. "Finska skällande fågelhunden 'Rehvi.'" In *Finska Kennelklubbens kalender och stambok*, 64–5. Helsingfors, Finland: Finska Kennelklubben, 1894.

Yhdeksännen yleisen Suomen maanviljelyskokouksen Wiipurissa 1887 luettelo. Helsinki: J. C. Frenckell'in ja Pojan Kirjapaino, 1887.

Zetterberg, Seppo. *Yhteisellä matkalla. VR 150 vuotta*. Helsinki: WSOY, 2011.

Part II
Interspecies Communities

6 Case Study of a Changing Human–Animal Relationship

Wild Rabbits in Britain from the Nineteenth Century to the Onset of Myxomatosis

John Martin

One of the most remarkable features of the European wild rabbit (*Oryctolagus cuniculus*) is its transformation in Britain from being a historically highly protected and esteemed species, in the mid-nineteenth century, to becoming officially classified as a major agricultural pest by the Second World War. In order to explain the changing and, at times, contradictory attitudes, this chapter will explore the early history of the wild rabbit, before focusing on the behavior of rabbits and their influence on British society during the nineteenth and twentieth centuries, when its numbers were increasing rapidly. It will also evaluate the reasons why, by the 1930s, the rabbit was one of Britain's most common species of mammals, causing the state to implement an extensive, multifaceted campaign to control wild rabbits during the Second World War. It will also explore the impact of myxomatosis, the highly infectious, subcutaneous viral pox, which devastated the rabbit population in the 1950s.[1] This decline constitutes one of the most remarkable ecological changes to have taken place in any European species of mammal or bird.

Britain, along with the Scandinavian countries, benefit from a temperate climate as a result of the Gulf stream. They also have a similar mode of agricultural production, which favors the proliferation of rabbits. The changing role of the wild rabbit, and its relationship with wider society, has been largely ignored by the academic community in Scandinavia as well as in Britain. There are exceptions, however, the most important in Britain being the accounts compiled by John Sheail.[2] Of more recent origin is Peter Bartrip's *Myxomatosis: The History of Pest Control*, which focused primarily on the history of the disease rather than its changing relationship with wider society.[3] Utilizing a wide range of primary sources, including official reports, correspondence and the records of the Oxford University Research Institute, the Bureau of Animal Populations (BAP), this case study explores the changing relationship between wild rabbits and humans since the nineteenth century up to and including the onset of myxomatosis in the 1950s.

The Long-Term Increase in the Rabbit Population

Following the initial importation of rabbits as a captive species in about the twelfth century, the animals were kept in specially constructed warrens or coneygarths, often enclosed by a fairly substantive moat or ditch filled with water to prevent them from escaping, and were protected by warreners.[4] Rabbits were being farmed both for their meat and fur, which was a highly esteemed, ornate clothing accessory. Rabbits, including the growing numbers which had escaped from warrens, were protected by legislation.

The late eighteenth and early nineteenth centuries provided a golden age for the increase in the numbers of wild rabbits. This expansion reflected a number of key changes which were taking place in the British economy and society. First and foremost were the effects of the Agricultural Revolution of the late eighteenth and early nineteenth centuries, when one of the unintended consequences of new crop rotations was the increasing availability of winter-sown cereals and improved grasses. These provided a more regular and plentiful supply of forage throughout the year, not only for farmed livestock but also for rabbits.[5]

Rabbits, being both promiscuous and prolific breeders, were ideally suited to be able to capitalize on the increase in their natural food supply. Unlike the vast majority of other mammals, the female rabbit does not have a regular oestrous cycle, with regular periods of heat, during which ovulation occurs. Once sexually mature, the doe can theoretically be presented to the buck at any time except during gestation but, in practice, she does not always accept the male. Instead, ovulation in the female is brought on by mating. More importantly, they have a short gestation period with a high fecundity rate (number of kindlings or young per unit of time) and a high proficiency (number of young per kindling). Consequently, in the absence of effective predator control in the form of animals, birds of prey and humans, rabbits, along with rats and mice, were able to increase faster than most other mammals. Those born in early spring, for example, would produce their own litters before the end of the same year.

Another factor favoring the rabbit was the prevalence of game shooting and conservation. During the late nineteenth century, it was the country's most rapidly expanding and popular rural sport, with new participants, almost exclusively members of the landowning classes or members of the "nouveaux riches," considerably outnumbering those who took up fox hunting.[6] Sharing the same natural predators as the pheasant and partridge, hares and rabbits benefited from the spread and intensification of game preservation. Vermin control on shooting estates made strenuous efforts to eliminate potential predators of game birds, such as weasels, stoats, and birds of prey, which had helped to keep the rabbit and rat population in check.

Rabbits, along with hares, which were classified as ground game, were protected by legislation, which preserved them from the ravages of poachers. The 1831 Game Act necessitated the purchase of a game licence and also imposed draconian penalties, including transportation, on those found guilty of poaching. The increase in the rabbit population, and the concomitant damage imposed on growing cereal crops and grassland, was an issue of serious concern for the many tenant farmers.[7] It was not until the landmark legislation of the Ground Game Act 1880, which conferred the right to kill rabbits and hares on both owners of land and tenant farmers, that the latter were finally legally able to kill rabbits and hares.

Even then this was a permissive Act, which landlords actively discouraged their tenant farmers from implementing strictly according to the letter of the law.[8]

Rural society had an ambivalent attitude to rabbits, which varied according to whether you were a member of the landowning classes and regarded them as part of your own private property, or whether you regarded them as a pest or common property, a source of meat which non-landowners were entitled to kill. Poaching of game birds, hares and rabbits, the latter probably accounting for nearly two-thirds of the poacher's haul, was still widespread even on those estates under the jurisdiction of gamekeepers. Between 1895 and 1899, convictions for poaching averaged 7838 a year, that is, over one conviction for every ten who possessed a game licence. It was undertaken not only by the poor and the unemployed but also, as Jones' research has shown, by miners, farmers, millers, butchers and even solicitors who enjoyed the thrill of engaging in an illicit activity, or to put one over the local squire.[9]

During the latter part of the nineteenth century there was a rapidly growing demand for wild rabbits both for the meat and fur. Indeed this demand was so great that the agricultural writer Richard Jefferies argued that the open and sandy heaths, on which it was difficult to grow arable crops satisfactorily, could once more be used for rabbit preserves.[10] While few farmers regarded rabbits as an agricultural crop which was comparable with conventional forms of crop production, commercial trapping of rabbits was widespread. Professional rabbit catchers, however, operated primarily during the winter and early spring, when the demand for rabbit meat and the quality of pelts was at its highest. Their strategy was to catch the harvestable surplus of the population while ensuring that a viable breeding stock remained.

Trapping was considerably cheaper and more efficient than shooting. Rabbits acquired in this manner commanded higher prices since they were not contaminated with lead shot which was suffused with blood and made them less attractive for consumers. Moreover, their pelts, free from pellet holes, attracted premium prices from the furriers for fashioning gloves, trimmings, coats or linings.[11] The most popular method of capture was by using spring or gin traps, which had been developed in the late nineteenth century. One of the unintended consequences of trapping was that it also caught a disproportionate number of the rabbit's natural predators, in particular stoats and weasels. According to estimates provided by Lockley, between ten and fifteen percent of the daily catch was made up of other species.[12] As a result, trapping, instead of acting as a means of controlling or even reducing the rabbit population, further facilitated its increase. In turn, this led to rabbits causing even more damage to growing crops and grassland.[13]

The First World War and the Post-War Depression

Changes in land management, accompanied with the mass sale of estates, mainly to their tenant farmers, during the First World War and immediately afterwards, impacted on the rabbit population. The enforced reduction in food imports during the war led to a plowing up campaign and intensification of agriculture, which was initially less favorable for rabbits. Wartime meat shortages encouraged more people to eat rabbits, while the reduction in the number of gamekeepers provided greater opportunities for poaching to take place. The increase in the number of owner-occupiers gave them greater jurisdiction over the way they farmed, including the extent to which they controlled rabbits. This decline in number of rabbits was however short-lived. The resumption of the agricultural depression in the early 1920s encouraged a switch to low-input methods of extensive production.

One of the more obvious features of the depression was the continuing decline in the arable acreage which, by the mid 1930s, was considerably lower than at any time since the onset of the Agricultural Revolution in the mid-eighteenth century. This was more than offset by the increasing dereliction of the countryside, which favored the rabbit population. As the Bureau of Animal Populations noted in relation to coverts, previously well-kept, manicured rides, where the guns could stand, were frequently allowed to become overgrown, impenetrable thickets, making it increasingly difficult to flush out birds. On the heavier, more intractable soils, alternate husbandry or ley farming was gradually abandoned, resulting in many temporary pastures being allowed to "tumble down" to nondescript, permanent pasture or even scrubland. By the 1930s the wild rabbit population in Britain had reached plague proportions, with a population of between 60 and 100 million, amounting to two or three to every acre of agricultural land.

The population increase was further prompted by the fact that, by this time, for the urban population at least, rabbits were regarded as a poor substitute for other kinds of meat, including beef, lamb, pork and even chicken, which were readily available. Moreover, the demand for British rabbits was further undermined by increasing imports of low-priced frozen rabbits from Australia, which flooded the British market. With both the demand for rabbits for meat or fur in the doldrums, it was not surprising that the amount of agricultural damage they caused reached an all-time peak. The damage was conservatively estimated to be in the region of £20 million per year, the equivalent of nearly 5 percent of annual gross agricultural output.[14] The vast majority of farmers regarded this as an occupational hazard, where, rather like the weather, it was assumed that there was little they could do to alleviate the problem. As the well-known writer and regular contributor to *the Countryman* magazine S. J. B. Thorburn lamented, among farmers there was "a temptation to resign themselves to the inevitable."

At a time of low agricultural prices, rabbits did provide compensatory benefits in the form of renting out the shooting rights, or more usually farmers allowing their sons to have the rabbiting in lieu of formal wages.[15] Social norms meant that farmers' sons often "depended on rabbits for their pocket money."[16] From the sons' point of view, completely exterminating rabbits was neither desirable nor even possible given the methods they usually employed. As research by the Bureau of Animals populations revealed, casual shooting of the type which was undertaken, or ferreting during the autumn and winter months, had little effect on total numbers.[17] It was only the small number of progressive farmers concerned about the need to maximize agricultural productivity who regarded rabbits as a pest in need of control or even eradication. The most systematic attempt in this respect was undertaken by brothers A. J. and F. N. Hosier, who regarded rabbits as a pest, to be controlled, whereas other farmers tended to regard them as a way of supplementing their income. The Hosiers had increased the acreage of the land they farmed so that at peak it was more than 25,000 acres in Wiltshire and Hampshire. Not only did they erect rabbit-proof fencing around some of the land, but when this proved inadequate they continued to employ full-time rabbit catchers during the summer months, when there was little demand for rabbit meat, and bury the rabbits they had caught.[18]

The increase in the amount of damage they caused to agriculture encouraged officials to investigate the possibility of controlling rabbits through the use of *Myxomatosis cuniculi*. This lethal virus had originated in South America in 1898, before appearing in California in 1930. Then in 1933, at the invitation of the Australian authorities, Sir Charles Martin, the internationally renowned physiologist and pathologist, undertook a series of experiments on the virus and potential methods of transmission.[19] His investigations, conducted under the auspices of the Cambridge University Department of Experimental Pathology, revealed that in a small, enclosed paddock, the disease was 100 percent fatal to wild rabbits.[20] However, more extensive experiments with the virus on the rabbit infested island on Skokholm in Pembrokeshire proved ineffective.[21] Martin was so disappointed with the failure that he retired shortly afterwards, concluding that the virus was not a viable way of dealing with the common rabbit.

Qualms were expressed about the way these control methods resulted in an unacceptable level of suffering on the animals. In 1938, the Universities Federation for Animal Welfare undertook more field trials on the Skokholm rabbit population. These showed that traditional method of gassing using sulphur caused a painful death, which was deemed unacceptable on humanitarian grounds. Such concerns did not, however, encompass the idea that it was wrong to exterminate rabbits by gassing, merely that it could be more humanely undertaken by the use of calcium cyanide in the dust form, known as "Cyanogas."

The Impact of the Second World War: From Food Surpluses to Food Shortages

The outbreak of the Second World War heralded an unprecedented shift in the state's relationship with the agricultural sector and in turn attitudes to food production. Domestic food production had to be raised to compensate for reduced levels of imported feeding stuffs, which declined from 22 million tons in 1938 to 11 million tons by 1944.[22] The wartime aim was to implement a structural realignment of agricultural production by increasing the arable acreage via a state-coordinated ploughing up campaign. This would increase the production of carbohydrate foods, particularly wheat and potatoes, which were officially regarded as the staple components of a wartime diet. In contrast, the livestock sector, which was less efficient in terms of producing calories, was forced to contract.[23]

The newly established Ministry of Food took responsibility for developing a dietetically adequate rationing system that was based on the supply of calories of vegetable origin. The Ministry of Agriculture continued to be responsible for directing production on the farm. In order to coordinate the food production campaign at the local level, a War Agricultural Executive Committee (WAEC) was established for each county, with the task of implementing national policy objectives by encouraging farmers to carry out their directives. The terms of reference under which they had been established required them to take responsibility for the control of rabbits, rats, mice and other pests, which reduced agricultural productivity. In the early stages of the war, WAECs experienced administrative difficulties in coordinating the pest control campaign. The Rabbits Order 1939 gave WAECs new powers under Regulation 63 of the Defence Regulations to enter and take rabbits upon any land.[24] The Ministry of Agriculture was very circumspect about implementing this Order in practice.[25]

Their powers were subsequently extended,[26] and following the recommendation of the Bureau of Animal Populations (BAP), Rabbit Control Officers were appointed to coordinate the task.[27] During the initial stages of the war, it continued to research and develop more effective methods of rabbit control.[28] In effect, according to BAP, while the means of being able to effectively control rabbits were now well known, many farmers either lacked the technical expertise or more likely the commitment to implement the more draconian control methods such as gassing.[29]

The most popular gas used was "cymag," a derivative of cyanide, which had first been used in the 1930s. With the increased demand for cyanide gas, supplies were quickly exhausted resulting in a two-month waiting list.[30] This method was not always successful, particularly on dry, sandy soils that were not capable of retaining a lethal density of the gas in the burrow.[31] Gassing was also unpopular amongst the farming community as dead rabbits could not be recovered easily, and more importantly, there was widespread opposition to eating them.[32]

These control measures, coupled with the increased demand for rabbits for human consumption, led to a significant decline in the rabbit population during the Second World War. According to the official Crop Reporters, this had resulted in a very marked decline in the amount of agricultural damage they caused. The accuracy of their responses is questionable, however, and, according to officials, the survey had served its purpose and no further surveys were undertaken.[33] It should be noted that farmers had a vested interest in under-recording the extent of damage in case it incurred the wrath of their local WAEC for failing to deal with the problem. The debatable issues are not only the precise extent of the decline but the relative importance of the causal mechanisms.

The Ministry of Agriculture and the WAECs must take some of the credit for persuading farmers to address the rabbit problem, as they distributed a number of instructional pamphlets. These were prepared by the BAP, who also organized courses for Pest Control Officers.[34] Campaigns were initiated for the widespread destruction of rabbits by trapping, snaring and ferreting, followed by gassing in an effort to eradicate any surviving pockets of rabbits.[35] Individuals and groups under the jurisdiction of the WAECs also played their part, although not all wartime activities were beneficial.[36]

Other factors also contributed to the wartime decline in the rabbit population. Opportunities for poaching were increased by the 65 percent reduction in the number of gamekeepers during the war, which meant that estates were less well preserved than in the past.[37] In addition, there was a much greater willingness, even among the respectable shooting fraternity, to disregard conventional shooting etiquette, such as not shooting ground game. This even entailed engaging in what were once considered nefarious activities, including trapping and snaring, to secure a rabbit or a variety of other animals and birds to eat.[38] The shift was prompted by the wartime food rationing restrictions, which meant that by 1944, average edible meat consumption had declined from 109.6 lbs to 96.1 lbs per person.[39] These quantitative changes obscured an even more fundamental qualitative change in the type of meat available. During the war when consumers went to the butcher with their rationing coupons, they were allocated meat whose origin was not always specified, nor was it necessarily fresh.[40]

Wartime meat shortages encouraged a buoyant black market economy for fresh rabbits.[41] Subsequent research indicated that wild rabbit was the only meat that some countrymen ate during the war years.[42] The food production campaign of the Second World War period has conventionally been hailed as a period of significant increases in agricultural output and, by inference, productivity.[43] According to contemporary official estimates, the 40 percent reduction in pest damage, over half of which can be attributed to rabbits, was equivalent to at least one million tons of food and feeding stuffs.[44] This was not only a conservative estimate, but also makes it evident that the wartime decline in rabbit damage appears to be probably the single most important factor accounting for the wartime increase in agricultural output.[45] The contribution of rabbits to saving Britain from not only malnutrition but even possibly starvation merits more credit than it has so far received.

The Post-War Period and the Impact of Myxomatosis

In 1947 the Agriculture Act replaced the 1939 legislation, transferring the powers of the Rabbits Order 1939 to County Agricultural Executive Committees. The new legislation introduced grant aid to assist farmers and landowners with the costs incurred in controlling rabbits via the establishment of Pest Destruction Societies. However, lack of publicity meant that there was little awareness amongst farmers of either the financial assistance available, or the procedures to be followed in order to acquire it. As Sheail astutely noted, "[b]ehind the rhetoric of the bargain struck between government and industry which offered guarantees of economic stability in return for higher productivity, little had been absorbed as to the detail by which such technical and organisational aspirations might be achieved."[46]

The most important change was the introduction of the landmark Prevention of Damage by Pests Act 1949. Legislative changes had three interrelated functions: to transfer the responsibility of rabbit control from the agricultural committees to landowners and occupiers, to maintain infestation surveys, and to advise farmers on methods of reducing pest damage.[47] It is also evident that not all farmers fully appreciated the extent of damage they caused but continued to regard them as a means of supplementing their income by letting out the right to shoot rabbits or allowing their sons to catch rabbits in lieu of wages.[48]

Pressure on the rabbit population was intensified by the post-war meat shortages. By the late 1940s, when austere rationing controls were more stringent than at any time during the war, the demand for illicit fresh meat including rabbits was at a premium, even if it was acquired as a result of clandestine activities. According to Tom Shakesheff's research into poachers, the immediate post-war period was best remembered for the opportunities it provided for the illicit taking of game and rabbits.[49]

Given these pressures, rabbit numbers and the damage they inflicted on agriculture were at considerably lower levels than those prevailing prior to the Second World War. Measuring the magnitude of these parameters is problematic, with estimates varying from the widely quoted figure of £50 million per year cited by the Myxomatosis Advisory Committee, to less than £20 million per year.[50] What is clear, though, is that in the autumn of 1953, pestilence brought about by myxomatosis devastated the rabbit population. The circumstances surrounding the subsequent localized outbreak at Bough Beech near Edenbridge, Kent, in October 1953, are still shrouded in mystery. However, the impact of the disease is not in dispute with mortality rates in excess of 99.9 percent being initially recorded in some localities.[51]

By the end of the year following the initial outbreak, myxomatosis had been reported from eight localities in four southeastern counties. It spread relatively slowly, with the average advance being about 1.5 miles a month. Its spread possibly was facilitated by farmers releasing diseased rabbits on their land.[52] By 1957 the disease was documented in thirty-four counties, while six years later there was only one county in England and one county in Wales without any known cases of the disease.[53] Myxomatosis has been widely acknowledged as the main causal factor accounting for the decline in the British rabbit population.[54] For the first time farmers began to fully appreciate the extent to which rabbits were responsible for reducing crop yields and, in turn, their profitability. Indeed, the highly respected naturalist, Robin Lockley, was ever rash enough to claim that 1955 would constitute "zero hour for the rabbit" with numbers he suggested would be lower than at any time since the eleventh century.[55]

To government officials, struggling to increase agricultural output and productivity to cope with the continuing post-war food shortages, a rabbitless countryside appeared a highly desirable policy objective. In an effort to ensure complete eradication, the government established mandatory Rabbit Clearance Societies (RCAs) in a number of areas. In spite of the number of RCAs increasing to 717 by 1964, they only covered 23 percent of farm and woodland in England and Wales.[56] Once more, not all farmers endorsed the need for such draconian control methods, allowing a few rabbits to survive as they were regarded as an integral part of the countryside, as well as offering the possibility of shooting a few for either the pot or a cash sale.

Even in areas under the jurisdiction of RCAs, complete eradication of the prolific and promiscuous rabbit proved elusive. The rapid recovery of the rabbit population was, however, constrained by the ongoing state-directed modernization of agriculture. These encompassed the grubbing up of hedgerows and scrub clearance coupled with a multitude of policies intended to encourage farmers to regard their activities as a business activity subject to profit and loss accounting rather than a way of life. These developments were more than offset by a number of changes which favored the revival of the rabbit population. The most important of these was when the previously single highly virulent strain of myxomatosis was gradually replaced by a multitude of strains differing widely in virulence.[57]

Since the 1950s, wild rabbit numbers have continued to increase, and in a number of areas have reached plague proportions, once more being responsible for considerable damage to crops and grassland. The unprecedented expansion of game shooting as a result of the revolutionary changes that have taken place in the artificial rearing of game birds since the 1950s has ensured that wild rabbits are no longer as important for game shooters as in the past.[58] In a similar way, the increase in the production of man-made artificial materials and alternative sources of natural fur has led to the collapse in the demand for rabbit fur.

Conclusion

The period from the nineteenth century to the mid-1950s witnessed a fundamental reappraisal in the state's approach to the wild rabbit. By the 1930s, rabbits, which were initially regarded as the exclusive perks of the landowning and sporting fraternity, protected by legislation, had reached plague proportions. They were having a significant impact on agricultural production, the extent of which was not fully appreciated until the late 1930s.

The decline in the rabbit population during the Second World War and immediate post-war period played a major role in accounting for the increase in agricultural output. The success of the campaign to reduce the rabbit population reflected not only the growing recognition of the damage they inflicted on agriculture, but also the fact that they were valuable as a source of meat for a population engulfed by austerity and stringent rationing controls. The most dramatic reduction in rabbit numbers was, however, the result of the spread of myxomatosis in the early 1950s, which possibly provided an opportunity to completely eradicate the mammal from Britain. Their subsequent survival and recovery is indicative of not only the ability to continue to prosper under a wide variety of conditions, but also the way rural inhabitants continued to favor its survival. In terms of human relationships with wild rabbits, while they have clearly been transformed since the nineteenth century, it is also possible to identify considerable signs of continuity.

Notes

1 Armour and Thompson, "Spread of Myxomatosis in the First Outbreak in Great Britain," 511–8.
2 Sheail, *Rabbits and Their History*; Sheail, "Wartime Rodent-Control."
3 Bartrip, *Myxomatosis*.

4 Sheail, "Changes in the Supply of Wild Rabbits," 175.
5 Ibid.
6 Martin, "The Transformation of Lowland Game Shooting," 1143.
7 Jefferies, "Rabbits as Food," 509.
8 For a more detailed analysis, see Sheail, *Rabbits and Their History*, 74–80.
9 Jones, "The Poacher," 825–60. See also Haggard, *I Walked by Night*, for an in-depth account of an unnamed poacher.
10 Jefferies, "Rabbits as Food," 509.
11 Campbell, "The Rabbit Problem," 287.
12 Lockley, *The Private Life of the Rabbit*, 126.
13 Tansley and Adamson, "Studies in the Vegetation of the English Chalk," 177–223.
14 Ministry of Agriculture, estimation made by Mr R. Genover in 1941. See National Archives MAF44/33.
15 Thorburn, "The Rabbit Racket," 262.
16 Ibid.
17 The Experimental Research Institute at Oxford University, Bureau of Animal Populations (hereafter BAP), Report Number 2.
18 Hosier and Hosier, *Hosier's Farming System*, 105.
19 Martin, *Observations and Experiments with Myxomatosis Cuniculi*, 10–38.
20 Lockley, "Some Experiments in Rabbit Control," 768.
21 Martin, *Observations and Experiments with Myxomatosis Cuniculi*, 10–38.
22 Hammond, *Food*, 5.
23 Murray, *Agriculture*, 375; Martin, *The Development of Modern Agriculture*, 51–2.
24 The 1939 Rabbits Control Order transferred powers given to the Local Authorities to the War Agricultural Executive Committees under Section 63.
25 Circular letter to County WAECs in England and Wales, October 13, 1939, WAC1151, Serial No. 41. See also BAP, Report No. 12, 1–10.
26 Ministry's Circular letter, January 10, 1940 (W.A.C.), Serial No. 99.
27 BAP, Report No. 2, 1.
28 Middleton, *The Control and Extermination of Wild Rabbits*; Crowcroft, *Elton's Ecologists*, 23–6; Sheail, "Wartime Rodent-Control," 59.
29 BAP, Report No. 21A, 2; BAP, Report No. 21B, 8.
30 BAP, Report No. 21B, 8.
31 Ministry of Agriculture, "The Control of Rabbits," 588; Martin, "The Impact of Government Intervention," 239–61.
32 BAP, Report No. 2, 3.
33 MAF 44/33; *Farmers Weekly*, December 27, 1940, 2.
34 BAP, Report No. 21C, 2.
35 BAP, Report No. 2.
36 Martin, *The Great Shoots*, 134.
37 Martin, "The Impact of Government Intervention," 240–2.
38 "Scolopax," "Rough Shoot Wanted," 133.
39 Hammond, *Food*, 391.
40 *The Field*, February 14, 1942, 181; *The Farmer and Stockbreeder*, April 16, 1940, 893.
41 Martin, "The Impact of Government Intervention," 254.
42 Davies, *The Wartime Kitchen and Garden*, 183; Lockley, *The Private Life of the Rabbit*, 129.
43 Murray, *Agriculture*.
44 Sheail, "Wartime Rodent-Control," 57
45 BAP, Report No. 21A, 2.
46 Sheail, "Wartime Rodent-Control," 66.
47 *Prevention of Damage by Pests Act* (1949).
48 Lillywhite, personal correspondence with the author.

49 Shakesheff, "For Profit and Pleasure," 5.
50 Bartrip, *Myxomatosis*, 134–5.
51 Armour and Thompson, "Spread of Myxomatosis in the First Outbreak in Great Britain," 511–8.
52 Mead-Briggs, "Rabbits and Myxomatosis," 197.
53 Southern, "Myxomatosis and the Balance of Nature," 247–8.
54 Mead-Briggs, "Rabbits and Myxomatosis," 197.
55 Lockley, "Myxomatosis Factual Survey," 1174–5; Southern, "If Rabbits Should Disappear," 1024–6.
56 Bartrip, *Myxomatosis*, 120.
57 Ross and Saunders, "Changes in the Virulence of Myxoma Strains in Britain," 113–7.
58 Martin, "The Transformation of Lowland Game Shooting," 1149.

Bibliography

Archival Sources

The Experimental Research Institute at Oxford University. Bureau of Animal Populations (BAP), Report No. 2. Recommendations for the Organisation and Technique of Rabbit Control through the Agricultural Committees during the Spring of 1940 (1940), 3. A.R.C4565/A.

BAP, Report No. 12. An Enquiry into County Organisations and Procedure for Dealing with Rats and Rabbits in England and Wales (1940).

BAP, Report No. 21A. Report by the Director upon Progress in Research during 1941–42 (1943).

BAP, Report No. 21B. Report by the Director on Progress in Research during 1943 (1944).

BAP, Report No. 21C. Report by the Director on Progress in Research during 1944 (1945).

Ministry of Agriculture, NTA, MAF44/33, LP3864 (1941).

W.A.C., Circular Letter to County WAECs in England and Wales, October 13, 1939, WAC1151, Serial No. 41.

W.A.C., Ministry's Circular Letter, January 10, 1940 (W.A.C,), Serial No. 99.

Newspapers and Magazines

Farmers Weekly, 1940.

The Farmer and Stockbreeder, 1940.

The Field, 1942.

Literature

Armour, C. J., and H. V. Thompson. "Spread of Myxomatosis in the First Outbreak in Great Britain." *Annals of Applied Biology 43* (1955): 511–8.

Bartrip, Peter W. J. *Myxomatosis: A History of Pest Control and the Rabbit.* London: I. B. Tauris, 2008.

Campbell, A. V. "The Rabbit Problem." *Journal of the Ministry of Agriculture* 46 (1939): 286–9.

Crowcroft, P. A. *Elton's Ecologists: A History of the Bureau of Animal Populations.* Chicago: University of Chicago Press, 1991.

Davies, Jennifer. *The Wartime Kitchen and Garden.* London: BBC Books, 1993.

Haggard, Lilia Rider, ed. *I Walked by Night: Being the Life and History of the King of the Norfolk Poachers*. London: Nicholson and Watson, 1935.

Hammond, Richard J. *Food: Volume 1: The Growth of Policy*. London: H.M.S.O., 1951.

Hosier, A. J., and F. N. Hosier. *Hosier's Farming System*. London: Crosby Lockwood, 1951.
Jefferies, R. "Rabbits as Food." *Farmers Journal* 6 (1877): 509.
Jones, D. J. V. "The Poacher: A Study in Victorian Crime and Protest." *Historical Journal* 22 (1970): 825–60.
Lillywhite, R. (the wartime owner of the "The Wilts" game farm, Andover, Hampshire). Personal correspondence with the author, May 13, 1987.
Lockley, Ronald M. "Some Experiments in Rabbit Control." *Nature* 145 (1940): 767–9.
Lockley, Ronald M. "Myxomatosis Factual Survey." *Field* (December, 1954): 1174–5.
Lockley, Ronald M. *The Private Life of the Rabbit: An Account of the Life History and Social Behaviour of the Wild Rabbit*. Newton Abbot: Reader's Union, 1976.
Martin, Brian P. *The Great Shoots: Britain's Premier Sporting Estates*. London: David & Charles, 1987.
Martin, C. J. *Observations and Experiments with Myxomatosis Cuniculi to Ascertain the Suitability of the Virus to Control the Rabbit Population*. Report of the Institute of Animal Pathology, University of Cambridge 4 (1934–45): 10–38.
Martin, John. "The Transformation of Lowland Game Shooting in England and Wales in the Twentieth Century: The Neglected Metamorphosis." *The International Journal of the History of Sport* 29 (2012): 1141–58.
Martin, John. "The Impact of Government Intervention on Agricultural Productivity in England and Wales 1939–45." PhD diss., University of Reading, 1992.
Martin, John. *The Development of Modern Agriculture: British Farming since 1931*. Basingstoke: Palgrave Macmillan, 2000.
Mead-Briggs, A.R. "Rabbits and Myxomatosis." *Agriculture* 73 (1966): 196–9.
Middleton, A.D. *The Control and Extermination of Wild Rabbits*. Oxford: Bureau of Animal Populations, 1940.
Ministry of Agriculture. "The Control of Rabbits." In *Agriculture*. London: MAFF, 1936.
Murray, K. A. H. *Agriculture*. London: H. M. S. O., 1955.
Prevention of Damage by Pests Act (1949). Accessed June 18, 2016. www.legislation.gov.uk/ukpga/Geo6/12-13-14/55/contents.
Ross, J., and M. F. Saunders. "Changes in the Virulence of Myxoma Strains in Britain." *Epidemiology and Infection* 98 (1978): 113–7.
"Scolopax." "Rough Shoot Wanted." *The Field* 7 (August 1943): 135.
Shakesheff, Timothy. "For Profit and Pleasure: Poaching in Post War Herefordshire, 1945–52." Paper presented at the Social History Conference, Belfast, November 21, 2003.
Sheail, John. "Changes in the Supply of Wild Rabbits." *Agricultural History Review* 19 (1971): 175–7.
Sheail, John. *Rabbits and Their History*. Newton Abbot: David & Charles, 1971.
Sheail, John. "Wartime Rodent-Control in England and Wales." In *The Frontline of Freedom: British Farming in the Second World War*, edited by Brian Short, Charles Watkins and John Martin, 55–66. Exeter: British Agricultural History Society, 2007.
Southern, H. N. "A Britain without Rabbits." *Discovery* 16 (1955): 186–9.
Southern, H. N. "Myxomatosis and the Balance of Nature." *The Listener*, February 16, 1956.
Tansley, A. G., and R. S. Adamson. "Studies in the Vegetation of the English Chalk, No. 3: The Chalk Grasslands of the Hampshire-Sussex Border." *Journal of Ecology* 13 (1925): 177–223.
Thomas, A. S. "Changes in Vegetation since the Advent of Myxomatosisis." *Journal of Ecology* 48 (1960): 287–306.
Thorburn, S. J. B. "The Rabbit Racket." *The Countryman* 30 (1944): 262.

7 Does a Dead Wild Animal Have Agency?

The White-Tailed Eagle as a Catalyst for an Ideational Revolution in Finland

Tuomas Räsänen

Introduction

By the 1970s the Finnish population of the White-Tailed Eagle (*Haliaeetus albicilla*) had collapsed to a few dozen individuals, whereas only a century earlier the species had been common in the archipelagic environments all along the Finnish coast. At the dawn of the twentieth century, the population was already in freefall. The White-Tailed Eagle, along with other predators, was considered to be a noxious animal and was hunted with bounties being used as incentives. The remnants of the eagle population survived in only three separate enclaves: two in the southwest corner of Finland (on the Åland Islands and in the Turku Archipelago) and the third in the Kvarken area on the west coast.[1]

The decline of the eagle population in the Northern Baltic Sea in the first part of the twentieth century did not go unnoticed. In the 1920s, the Swedish ornithologist and nature writer Bengt Berg bemoaned in his book *De sista örnarna* how there were only eighteen breeding couples remaining in the whole of Sweden, which undeniably meant that the White-Tailed Eagle was about to disappear from the country's fauna.[2] His pleas to save the eagles for the present and future generations were answered almost immediately. Sweden protected White-Tailed Eagles in 1924, only a year after the publication of *De sista örnarna*. In Finland, a preservation order was first put into effect in 1924 on the Åland Islands and in the rest of the country two years later.[3] Unfortunately, from the eagles' point of view, the status of being a preserved species did not really help them, since human hatred towards them continued and illegal killings remained ubiquitous.

The plight of White-Tailed Eagles finally received nationwide coverage in the late 1960s, when it was publicized that breeding problems, caused by toxic chemicals, had led to the population being on the verge of extinction in the northern Baltic Sea region. This coincided with a growing concern over the Baltic Sea marine environment. However, I argue that the will to protect the White-Tailed Eagle was not merely a side effect of increased environmental awareness. Instead, the White-Tailed Eagle played a part in making environmental attitudes among humans more sensitive to the deterioration of the marine environment by acting as a crucial node within a network, which not only consisted of humans and eagles, but also Baltic fish, different bacteria and synthetic chemicals, and functioned as indicators of change in the entire network. Thus, the White-Tailed Eagle offers an excellent case in point when examining multispecies interaction, which had noteworthy consequences for Finnish society. Eagle bodies (dead and alive) were points of convergence, in which the science of anthropogenic environmental change and the conservation of the natural world became entangled and produced emergent ideas about the relationship between humans and the marine environment. Consequently, eagles and the marine environment gained an elevated status in Finnish environmental thought. Thus, although the main focus of this chapter is to examine the change in the relationship between humans and the White-Tailed Eagle, it also sheds light on the evolution of marine consciousness in Finland in the 1960s and 1970s.

Animal agency is integral to the question of the eagles' role in Finnish society. According to Hans-Johann Glock, philosophers have often denied animal agency, because animals were deemed to be devoid of intentional behavior and being able to use complex symbolic language.[4] At the other end of the spectrum, actor-network theorists have been inclined to accept the agency of all objects, including microbes, rocks, seat belts and ships.[5] A middle ground position is adopted in this chapter. Current knowledge of animal behavior suggests that the difference between humans and many "higher animals" relates to degree rather than a categorical difference in kind, since many animal species use symbolic language and exercise various aspects of social behavior. They also have their own perspective on the world, which they can vary according to changing circumstances. "On the other hand," the notion of agency in the actor-network theory is based on a conceptual muddle, which, interestingly, actor-network theorists do not often even attempt to clarify.[6] As sociologists Bob Carter and Nickie Charles have pointed out, if any kind of object, irrespective of whether it is capable of performing independent acts, is an agent, then the position taken by actor-network theorists indicates that agency equates to being.[7] The actor-network theory, in other words, confuses agency with effect. Seat belts affect the lives of individuals and they also affect human social relations and morality, but it is the actions of inventors, legislators and users that actually transform the relevant social networks.

Hence, non-living beings are best viewed not as actors or agents, but as mere factors (or intermediaries in Latour's terminology). These factors are genuine members of social networks, since they possess causal power, which originates from their relationship with the network's other constituents. They can, for example, enable, govern or deter human actions. Yet, non-living objects act in a similar manner in each situation, while agents, whether human or "higher animals," can modify their behavior according to changing circumstances. They can also choose between alternatives, as well as benefit from their choices, whether they are intentional or not, or alternatively suffer as a result of these decisions. In so doing, animals can autonomously change entire social networks through their actions.[8] This also implies that animals are important agents of change in human history.

It may not be possible, nor is it important for the purpose of this chapter, to judge which animal species has agency, an autonomous capacity to act within and adjust their social networks. What matters is that White-Tailed Eagles clearly have agency, which has an effect on humans. Eagles, for example, reduce fish stocks, and their behavior forces domestic animals to be sheltered. However, as will be demonstrated below, the closest and most influential encounters between humans and eagles often occurred when zoologists and conservationists found and studied dead eagles. This raises the question of the temporal limits of animal agency (or, for that matter, of human agency). Are animals (including humans) still agential beings – or are they still actors – even after they cease to act when they die? Although the main concern of this chapter is to examine the changing relations between humans and the White-Tailed Eagle in Finland in the 1960s and 1970s, I will also seek answers to the problem of the agency of dead animals.

Fish Eaten and Farm Animals Stolen

The White-Tailed Eagle is the largest raptor in Finland. It used to breed in all Finnish coastal areas and in several large inland lakes. Contrary to long-held perceptions, the White-Tailed Eagle is a rather mediocre killer. In Finland, the species has a seasonal diet, as it tries to take advantage of the most convenient and accessible catches. The most important prey for the eagle are fish, such as pike, perch and roach. In the spring, it also preys on different bird species, of which waterfowl varieties are the most common. Occasionally, White-Tailed Eagles eat small mammals, such as weasels, hares and squirrels. In winter, when the sea is mostly frozen and the majority of birds have migrated, the White-Tailed Eagle often has to make do with scavenging carcasses. The eagles use different tactics when trying to catch prey. Sometimes they fly above the archipelago without any obvious intention to hunt. By so doing, it may lead to a weak bird becoming separated from a fleeing flock. The eagle then focuses on hunting this individual (thus acting as a selective force that improves the health of bird populations). Sometimes the eagles try to surprise prey by flying just above the tree line or by waiting patiently and inconspicuously on a tree for hours.[9]

From the human perspective, White-Tailed Eagles were viewed as traditional competitors for food, since they were (falsely) believed to have a detrimental impact on fish and waterfowl populations. Consequently, the elimination of White-tailed eagles was considered to be the most important goal in game management practices in the archipelago areas.[10] In addition to competition over food, human hatred towards White-Tailed Eagles stemmed from a fear that the birds of prey plundered domestic animals and even children. While there have been verified cases of White-Tailed Eagles catching lambs and dogs, the killing of domestic animals has been rare. In most cases, the White-Tailed Eagle was blamed for killing lambs, although the real killers were more often golden eagles.[11] As for the stories about White-Tailed Eagles stealing children, they are completely imaginary.

In human–animal relations, the latter are never just material beings as they also carry meanings that derive from this very relationship. Although fear of White-tailed eagles has been based on erroneous conceptions of the eagles' way of life, there is no denying that this emotion was genuine. The White-Tailed Eagle used to embody nature threatening civilization. In other words, its agency contravened the interests of humans, who acted accordingly to secure their own well-being. On the other hand, the eagles also developed survival strategies towards human aggression, from their own perspective, by changing their behavioral traits in a manner, which scholars have sometimes referred to as phenotype plasticity.[12] First, the eagle population retreated to the outer archipelago areas, where the human population was sparse and visitors rare. Second, eagles, which had earlier nested at the top of high trees, from where they could see far and wide, now sought to hide in the inner part of their home island and often chose lower trees for nesting. Third, eagle pairs used several nests in turn, which improved their successful breeding rates.[13]

It is possible that two distinctive developments took place as the eagles adapted to changing circumstances. On the one hand, it seems that the change in the eagles' behavior at least partly resulted from unconscious anthropogenic evolution. The killing of eagles functioned as a selective factor, which favored specimens that feared humans the most, while the most courageous specimens were killed. On the other hand, individual eagles may also have learned to avoid humans as best as they could.[14] The discovery of shot eagles and destroyed nests proved that even if a learning process took place among the birds of prey, it was not always enough to ensure their survival. Even worse, though, was a new chemical threat, about which the eagles knew nothing and which they could not escape, because it had penetrated everywhere in their food cycle.

Toxic Eagle Bodies

The toxic bombardment of the environment by humans accelerated after the Second World War. Modern society invariably required the development of new substances for many industrial purposes as well as for eradicating pests. Without proper purification systems and knowledge of the material flows in ecosystems, huge amounts of toxic substances were released into the environment. These substances often found their way into the sea. In the late 1960s, it was discovered that toxins had travelled via oceanic and atmospheric currents all the way to the circumpolar areas.[15] The Baltic Sea was particularly vulnerable to environmental contamination, because of the poor water exchange with the Atlantic Ocean and relatively low water mass.

In 1966, a study by Kurt Henriksson, Eeva Karppanen and Matti Helminen confirmed that eagles in Finland contained high concentrations of mercury.[16] Interestingly, however, the notion of the pureness of the Finnish environment did not die immediately, as it was upheld by an inventive theory that was advocated by some leading zoologists. According to this theory, White-Tailed Eagles were exposed to mercury in their bodies while wintering in ice-free areas of the southern Baltic Sea. This was a logical argument to explain the toxicity in the bodies of the eagles, as while mercury had been extensively used in Finland, the chemical had mainly been utilized in a relatively harmless form. This was in contrast to Sweden, for example, where industry had used lethal methylmercury.[17] However, this did not explain why old eagles that remained in their territories for the entire year also had high concentrations of toxic substances in their bodies.

It is not difficult to see the similarity between the arguments based on the toxins being an external source and the way Finnish people have imagined their country. After the burgeoning of romanticism in the nineteenth century, the Finnish intelligentsia began to portray the Finnish interior as a place where hardworking poor folk toiled to survive in the harsh northern climate, which separated them from the urbanized and civilized centers of commerce and manufacture in Europe and European-like cities on the Finnish coast. Finnish identity was partly based on the idea that European nations were rich in productive agricultural land and industry, but poor in the quality of their nature. In contrast, Finland still had plenty of unspoiled nature for its unbesmirched natives to use and revere.[18] In the 1960s, Finland was still seen by many Finns as an isolated edenic island set amidst a rapidly deteriorating global environment, despite being in the throes of rapid urbanization and industrialization. Mercury was a serious environmental toxin, but Finns had nothing to worry about it, since industrial pollution was the problem of manufacturing nations in the south. This Finnish exceptionalism can also be seen in the views of Finns towards other toxic chemicals. For many years Finnish scientists and naturalists downplayed the dangers of DDT to humans and the environment, despite the growing concern in other industrial nations. They argued that the need and use of pesticides in the cold climate of the North was much lower than in almost any other European country, and therefore posed less risk to Finland.[19] Northern nature was not only pure by definition, but also safeguarded Finnish land and watercourses from the harmful side effects of industrial society.[20]

It did not take long, however, to discover that Finland was not immune to the problems faced by other European states. Some eagles may have been contaminated with mercury from the southern Baltic Sea, but certainly not fish. When scientists measured alarmingly high mercury contents in fish caught from the Finnish sea area, it was impossible to deny that it was of domestic origin.[21] The contamination centered in watercourses downstream from industrial wood processing plants. Studies conducted by the Swedish chemists Arne Jernelöv and Sören Jensen soon proved that bacterial activity in water had transformed the mercury into methylmercury.[22] Because of the existence of extensive coniferous forests, which European countries at lower latitudes lacked, Finland's economy relied heavily on the wood processing industry. The large-scale wood processing industry in Finland and, to a lesser extent, the chloralkali industry, had used relatively harmless mercury, which then transformed into methylmercury at sea. From the seawater this methylmercury moved into the food chain from plankton to fish and finally accumulated in eagles in deadly quantities. Instead of shielding the environment from industrial contamination, northern ecosystems in Finland had become the hot spot of mercury pollution.

Jensen and his colleagues also published a study in 1969 in the journal *Nature*, which disclosed that Baltic fish contained more DDT and PCBs than fish caught in any other sea area in the world. In 1966, Jensen revealed that PCBs were environmentally toxic, by studying dead White-Tailed Eagles brought to the Swedish Museum of Natural History in Stockholm. These substances accumulated in the food chain in a similar manner to mercury, with eagles and other top predators being the most vulnerable to these toxins. The highest measurement from one individual White-Tailed Eagle specimen in Sweden showed that 3.6% and 1.6% of the weight of its body fat was composed of DDT and PCBs respectively.[23] Finnish scientists recorded concentrations that were almost as high. Because of this toxic cocktail, either adult specimens would die, as many already had, or they were consequently unable to reproduce viable nestlings. This latter effect was demonstrated by findings regarding the thinning of raptors' eggshells.[24] Either way, in the long run the result would be the same: the inevitable extinction of the White-Tailed Eagle in the northern Baltic Sea area.

The Symbol of the Contaminated Environment

The protection of endangered species had been a core feature of the conservationist agenda of the Finnish Association for Nature Conservancy (hereafter FANC) since its establishment in 1938. Yet, conservationists paid surprisingly little attention to the plight of the White-Tailed Eagle prior to the end of the 1960s. In 1963, for example, the White-Tailed Eagle was not mentioned in the summary of FANC activities on endangered species that was published in an anniversary issue of conservationists' *Suomen Luonto* magazine that marked 25 years of conservation work in Finland.[25] The actions of the conservationists concentrated almost solely on terrestrial ecosystems, such as the preservation of the last remaining free-running rapids and forest wilderness. In terms of endangered raptors, the focus of conservationists was on birds that populated remote wildernesses, such as the golden eagle and the peregrine falcon. This highlights the fact, as I have shown elsewhere, that conservationists at the time did not have a marine narrative. For conservationists and society as a whole, the sea barely existed except as a resource for fishermen and as a route for commerce.[26]

Among conservationists, ornithologist Göran Bergman was the first one to vocally champion eagles in the 1950s.[27] The real turning point, however, in the human–eagle relationship in Finland seems to have been the publication in 1967 of Teuvo Suominen's popular book *Lintujemme katoava aateli* (*Our Dying Noble Birds*).[28] By this time there was enough scientific evidence for Suominen to categorically dismiss persecution as being a principal threat to eagles. The majority of eagles that were examined had no visible injuries. Instead, their body tissue contained lethal doses of toxins.[29] At the same time environmental toxicology revealed the domestic origin of the toxic chemicals, thereby ensuring that concerns over the Baltic Sea environment and White-Tailed Eagles went hand in hand.

The White-Tailed Eagle was heading towards extinction in the Northern Baltic Sea area, because they occupied the top echelon in the marine food chain—alongside human beings. The nexus between eagles and humans was often explicitly acknowledged and exploited by conservationists, when trying to win support for their cause.[30] Efforts to elicit favorable sentiments towards the White-Tailed Eagle in Finland were greatly aided by news in the late 1960s from the other side of the world. In Minamata, a small Japanese fishing village, several thousand humans had fallen ill and many died after eating fish contaminated with methyl mercury. Although the content of this chemical was lower in Finnish fish samples, they were nonetheless far above the values that were considered safe by food safety authorities. The combined effect of the Minamata catastrophe and poisonous fish in the Baltic Sea was that ordinary people realized that the sea was in a state of crisis. Awareness of this problem also led to a popularization of the woeful plight of the White-Tailed Eagle population, which was being decimated by the effects of toxic chemicals. The sad situation faced by the eagles was perceived by many people in Finland as a warning sign to stop chemical pollution. The poisoned fish not only seemed to signal the imminent extinction of the White-Tailed Eagle from the northern Baltic Sea, but also warned of the real threat to human life, which made them both co-victims of industrial society.[31]

The notion of shared victimhood also translated into a perceived need among conservationists to reassess the hierarchy of birds of prey in Finland. In his book, Suominen discussed all the endangered raptors in Finnish nature. At the time Suominen was writing wilderness birds still received more attention in conservationists' minds. The emergence of the Baltic Sea environmental crisis meant that the previously undervalued White-Tailed Eagle secured growing status among raptors. The more eagles failed to breed, the more conservationists directed their efforts at their survival. In fact, based on the quantity of articles in *Suomen Luonto* magazine, the White-Tailed Eagle received more attention in the late 1960s and early 1970s than any other wild animal in Finland.

In the early 1970s, besides enlightening people about the dire situation faced by White-Tailed Eagles, conservationists also launched campaigns to protect the species. They for example sold posters, postcards and letter seals in order to raise money to buy and conserve peaceful nesting areas for eagle pairs, as well as guard individual nesting trees and compile population inventories. By far the most important method, however, was to feed eagles with clean food during the harsh winter months, when eagles were weakened by the toxic chemicals and suffered from a scarcity of food. These activities intensified in 1973, when FANC members established a WWF branch in Finland and *Operaatio Merikotka* (Operation White-tailed eagle), as the project was called, joined the WWF's international project for conserving the White-Tailed Eagle in northwestern Europe.[32]

When looking at these campaigns, they seem to manifest traditional features associated with conservation, whereby an emphasis was on preserving birds and their habitats. Yet, a bigger picture emerges when these campaigns are put in the context of contemporary trends in conservation ideology. As long as the sea was perceived as being pristine and an unthreatened part of nature, conservationists could have continued ignoring it. Along with chemical pollution, the Baltic Sea entered into the conservationists' psyche as a deadly human environment. Consequently, conservationists developed a marine narrative, which elevated the Baltic Sea and its archipelagoes into the canon of national landscapes, thus associating the sea with Finnish self-understanding.[33] The White-Tailed Sea Eagle became a majestic icon of this reconstructed, yet severely polluted nature. As a group of conservationists stated: "the dreadful state of the White-Tailed Eagle population mirrors, as one indicator, the threatened state of the whole archipelagic nature in this area."[34] Classical conservation methods mixed with modern environmentalism, and a grim situation of the eagle became intermingled with the faith of the sea and, indeed, of the human race, since "humans may run up against the same creeping death that they are spreading everywhere around themselves,"[35] as one anonymous conservationist put it. Posters depicting the eagle, for example, were not simply designed to raise money for the conservation of the eagles, but were as much about presenting the warning signs of the environmental crisis faced by humanity.

Eagles as Agents in a World of Environmental Crises

In their attempt to introduce animals as members of human politics, Sue Donaldson and Will Kymlicka have categorized animals according to their physical and mental closeness to humans. In this categorization, wild animals (such as the White-Tailed Eagle) are classed as "sovereign" animals, comparable to "foreigners" or "aliens" in human political relations. The Relational Agency of these foreigners is restricted to occasional interventions from the other side. Eagles, for example, may steal domestic animals and pets, which in Donaldson's and Kymlicka's categorization are referred to as "citizens." Humans, on their part, kill eagles and inhibit their ability to breed. The relationship is thus dominated by fear, although it must be said that eagles have a much greater reason to fear humans than vice versa.[36] However, in the world of shared victimhood, the relationship between humans and wild animals is much more blurred than the notion of foreigners or aliens assumes. For example, many wild animals have lost some of their fear of humans and have adjusted to urban life in recent decades,[37] which naturally changes their agency. As for the perspective of White-Tailed Eagles, a fear of humans has prevailed, but in the early 1970s this fear was more predominant than it is today. When conservationists fed eagles with carcasses, they noticed that the eagles always waited for hooded crows, ravens and seagulls to eat first before they felt comfortable enough to take their fill.[38]

The discovery of toxic pollution was one of the key elements in the transformation of Finnish conservationism into an environmental movement, which adopted the well-being of the human environment into the core of its agenda.[39] However, environmental ideas do not sprout in a vacuum, but are developed through engagement with the environment, including animals.[40] As I have demonstrated in this chapter, eagles were knowledge-brokers vis-à-vis the poisoned food chains in the Baltic Sea.[41] The new environmental awareness in Finnish society was, therefore, partly the result of the interagency of humans and eagles, which both ate Baltic fish in their diet. Eagles followed their species-typical behavior: they ate fish and waterfowl. Yet, this food now killed them, which led zoologists to study them and to find a huge amount of toxic chemicals in Baltic fish. In short, instead of being foreigners or aliens, eagles became members of the environmental community. From this position they facilitated change in human conceptions and action. Consequently, by the early 1970s, the most notorious toxic chemicals were banned in Finland and elsewhere around the Baltic Sea. The outlawing of these chemicals enabled more successful hatching among the eagles and the eagle population slowly started to recover.

The eagles that contributed to the findings of environmental toxicology were all dead: some had been hit by trains, others killed by a toxic overdose or pellets in their body. Irrespective of how and why they had arrived at the scientists' laboratory, all the birds were riddled with chemicals. Does this mean that the dead eagles had agency, as I asked at the beginning of this chapter? The intuitive answer would be no, since dead animals cannot be actors. Instead, they are similar to objects, such as rocks or seat belts, which may contribute to human history, but are unable to autonomously make alterations to human and non-human networks. There is a crucial difference, however, between a dead animal and a seat belt, for example, as the former was once an agent, while the latter has always been an inanimate object. When agency is understood as a relational concept this difference has potentially definitive implications.

Let me clarify my point through an example from human social relations. Dead people do not act any more than dead eagles, but from the perspective of his/her relatives, his/her past actions may still continue to have a tremendous impact on their lives. Some may even think that the deceased continue to live among them and dictate their actions (just think about religious teachers and their followers). In such a case, it is nothing but semantics to demark the boundaries of agentiality as only extending to the moment of death. To sum up, eagles do not display agency on their death. Yet, their species-typical behavior led to their death, which consequently also exposed the catastrophic state of the Baltic Sea to humans.

Thus, although the behavior of eagles remained more or less the same, dead eagles instigated a radical change in a way the agency of the survivors of the species were conceptualized. Hardly anyone looking at the graphic images disseminated by conservationists, for example, cannot help sensing the grandeur of an eagle in flight. This sense of awe even applied to the local people, who had an ambivalent relationship to eagles, as Bengt Berg's story of the old fisherman testifies. On the one hand, there was a deep-seated hatred, depicted by Berg in the fisherman's instinctive urge to shoot an eagle on sight. On the other hand, similarly to the same Berg's fisherman, many people felt awed by the majesty of a flying eagle.[42] It was just that the negative aspects of eagle agency meant more to them, even to those who cared at all (for there were obviously many who did not), and thus drove them to kill the beast despite its sublimity. When economic necessities in the late twentieth century no longer imposed similar priorities and "nature began to seem a less overwhelming opponent,"[43] many locals joined with conservationists and bird lovers in admiring the beauty of the eagle.

In a short period of time in the 1960s and 1970s the semiotic bearings of the eagle turned upside down. Instead of nature being perceived as threatening civilization, the former was being imperiled by the latter. Consequently, the White-Tailed Eagle was transformed from being a largely ignored species to a flagship animal, whose survival was a matter of great concern, not only for the whole conservation movement, but also an increasing number of ordinary Finns. What is more, the dead eagles, as indicators of an invisible toxic threat, made Finns realize the ontological likeness of eagles to humans. Eagles, for their part, have once again gained more trust in human beings. They no longer have to hide from human eyes, but instead many choose to situate their territory near human dwellings. After centuries of fear on both sides, humans and eagles have found, at least temporarily, tranquil and mutually beneficial ways of co-living in Finnish coastal areas.

Acknowledgements

This research was supported by a grant from the Academy of Finland (274573). I am grateful to prof. Sverker Sörlin and researchers at the KTH Higher Seminar for their comments and Dr. Robert Collis for correcting the language.

Notes

1 Suominen, *Lintujemme katoava aateli*, 68; Pohja-Mykrä, Vuorisalo and Mykrä, "Organized Persecution," 11.
2 Berg, *Viimeiset kotkat*, 60–1, 67–8 and passim.
3 Suominen, *Lintujemme katoava aateli*, 67–70.
4 Glock, "Animal Agency," 384.
5 Sayes, "Actor-Network Theory and Methodology," 136, 138.
6 As the actor-network theorist Edwin Sayes states: "the term 'agency' is almost empty of meaning for ANT." According to Sayes, the actor-network theory should not be understood as a theory of agency, but as a methodological framework guiding social scientific research. Thus, the greatest value of the actor-network theory lies in its postulation of the significance of non-human beings in human social life and human history. See, Sayes, "Actor-Network Theory and Methodology," 144.
7 Carter and Charles, "Animals, Agency and Resistance," 324.
8 Carter and Charles, "Animals, Agency and Resistance," 327–8, 334–6. See also Salo, "Uusiutuva energia ja energiajärjestelmän konsensus," 60, 64.
9 Koivusaari, Nuuja and Palokangas, *Uhattu merikotka*, 50–6.
10 Suominen, *Lintujemme katoava aateli*, 67.
11 Suominen, *Lintujemme katoava aateli*, 68, 70. See also, Koivusaari, Nuuja and Palokangas, *Uhattu merikotka*, 54.
12 For the concept of phenotype plasticity, see, for example, Foote and Gunnels, "Exploring Early Human-Animal Encounters," 208–9.
13 Suominen, *Lintujemme katoava aateli*, 70.
14 Torsten Stjernberg, in discussion with the author. It was unconscious in the sense that human beings were not aiming to change the genetic traits of the eagles. See Russell, *Evolutionary History*, 17–8.
15 See for example, Peterle, "DDT in Antarctic Snow," 620; Tatton and Ruzicka, "Organochlorine Pesticides in Antarctica," 346–8.
16 Henriksson, Karppanen and Helminen, "High Residue of Mercury," 41–3.
17 Henriksson, Karppanen and Helminen, "High Residue of Mercury," 42; Suominen, *Lintujemme katoava aateli*, 72–4; Räsänen, "Itämeren ympäristökriisi," 62–3.
18 Klinge, *Kaksi Suomea*, 97–129.
19 For a more extensive discussion see, Räsänen, "Converging Environmental Knowledge," 163–5.
20 This same notion of the purity of the Finnish environment can be seen in contemporary Finland, in domestic discussion regarding food safety. See, for example, Kotilainen, "Puhdasta, suomalaista, nationalistista lihaa," 49–50.
21 Häsänen and Sjöblom, "Kalojen elohopeapitoisuus Suomessa."
22 Jensen and Jernelöv, "Biological Methylation of Mercury," 753–4.
23 Jensen et al., "DDT and PCB in Marine Animals," 248–9; Jensen, "Report of a New Chemical Hazard," 612; Räsänen, "Converging Environmental Knowledge," 166–70.
24 For scientific studies on the thinning of raptors' eggshells in the Baltic Sea area and elsewhere, see Räsänen, "Itämeren ympäristökriisi," 93–4.
25 Mikola, "Neljännesvuosisata luonnonsuojelutyötä."
26 Räsänen, "Itämeren ympäristökriisi," 184–5, 262.

27 Bergman, "Merikotkatilanne."
28 See also Telkänranta, "Tiedotusvälineiden muuttuva maailmankuva," 223.
29 Suominen, *Lintujemme katoava aateli*, 72, 74.
30 See, for example, "Vielä biosideista"; Helminen, "Luonnon kemialliset saasteet I," 30; Henriksson, "Elohopea ja saaristomme eläimistö," 123; Joutsamo and Hedenström, "Turun saariston merikotkat vuonna 1972," 188.
31 See, for example, Andrejew, "Tuholaismyrkystä tullut tuholainen."
32 "Maailman Luonnon Säätiön Suomen Rahasto," 106–7; Hautala, "Kotkien talviruokinta"; Koivusaari, Nuuja and Palokangas, "Merenkurkun merikotkain perikato," 160; Jacobsson, "Att mata örnar," 62; Hallander, "Att utfodra rovfågel"; Helander, in discussion with the author.
33 Räsänen, "Itämeren ympäristökriisi," 190–1.
34 Koivusaari et al., "Suomen merikotkat v. 1973," 174.
35 "Vielä biosideista," 32.
36 Donaldson and Kymlicka, *Zoopolis*, 14.
37 See, for example, Wembridge, *Urban Mammals*.
38 Hautala, "Kotkien talviruokinta," 110.
39 Räsänen, "Converging Environmental Knowledge," 173–4.
40 Nash, "The Agency of Nature," 68–9.
41 See also "Kalaelohopea ja luonnonsuojelu," 18.
42 Berg, *Viimeiset kotkat*, 42.
43 Citation from Ritvo, *Noble Cows and Hybrid Zebras*, 205.

Bibliography

Andrejew, Kiti. "Tuholaismyrkystä tullut tuholainen." *Helsingin Sanomat*, July 27, 1969.

Berg, Bengt. *Viimeiset kotkat*. Translated by Eino Kärki. Porvoo, Finland: WSOY, 1930.

Bergman, Göran. "Merikotkatilanne." *Suomen Luonto* 23 (1964): 113–8.

Carter, Bob, and Nickie Charles. "Animals, Agency and Resistance." *Journal for the Theory of Social Behaviour* 43 (2013): 322–40.

Donaldson, Sue, and Will Kymlicka. *Zoopolis: A Political Theory of Animal Rights*. Oxford: Oxford University Press, 2011.

Foote, Nicola, and Charles W. Gunnels IV. "Exploring Early Human-Animal Encounters in the Galapagos Islands Using a Historical Zoology Approach." In *The Historical Animal*, edited by Susan Nance, 203–20. Syracuse, NY: Syracuse University Press, 2015.

Glock, Hans-Johann. "Animal Agency." In *A Companion to the Philosophy of Action*, edited by Timothy O'Connor and Constantine Sandis, 384–92. Chichester, UK: Wiley-Blackwell, 2010.

Hallander, Håkan. "Att utfodra rovfågel." *Sveriges natur* 59 (1968): 10–5.

Häsänen, Erkki, and Veikko Sjöblom. "Kalojen elohopeapitoisuus Suomessa vuonna 1967." *Suomen Kalatalous* 36 (1968): 1–24.

Hautala, Hannu. "Kotkien talviruokinta." *Suomen Luonto* 29 (1970): 108–13.

Helander, Björn (zoologist, Swedish Museum of Natural History). In discussion with the author. April 2015.

Helminen, Matti. "Biosidit ja petolintujen kohtalo." *Suomen Luonto* 25 (1966): 100–3.

Helminen, Matti. "Luonnon kemialliset saasteet I. Viljelysten ja vesistöjen elohopeaongelma." *Suomen Luonto* 27 (1967): 29–30.

Henriksson, Kurt. "Elohopea ja saaristomme eläimistö." *Suomen Luonto* 30 (1971): 123–4.

Henriksson, Kurt, Eeva Karppanen, and Matti Helminen. "High Residue of Mercury in Finnish White-tailed eagles." *Ornis Fennica* 43 (1966): 38–45.

Jacobsson, Torsten. "Att mata örnar." *Sveriges Natur* 54 (1963): 62–4.

Jensen, Sören. "Report of a New Chemical Hazard." *New Scientist* 32 (1966): 612.

Jensen, Sören, and Arne Jernelöv. "Biological Methylation of Mercury in Aquatic Organisms." *Nature* 223 (1969): 753–4.
Jensen, Sören, A.G. Johnels, Mats Olsson, and Gunnar Otterlind. "DDT and PCB in Marine Animals from Swedish Waters." *Nature* 224 (1969): 247–50.
Joutsamo, Esko, and Kaius Hedenström. "Turun saariston merikotkat vuonna 1972." *Suomen Luonto* 31 (1972): 188–9.
"Kalaelohopea ja luonnonsuojelu." *Suomen Luonto* 29 (1969): 14–20.
Klinge, Matti. *Kaksi Suomea*. Helsinki: Otava, 1982.
Koivusaari, Juhani, Ismo Nuuja, and Risto Palokangas. "Merenkurkun merikotkain perikato." *Suomen Luonto* 29 (1970): 158–61.
Koivusaari, Juhani, Ismo Nuuja, Risto Palokangas, Esko Joutsamo, Kaius Hedenström, Torsten Stjernberg, and Henrik Wallgren. "Suomen merikotkat v. 1973." *Suomen Luonto* 32 (1973): 174–7.
Koivusaari, Juhani, Ismo Nuuja, and Risto Palokangas. *Uhattu merikotka*. Jyväskylä: Gummerus, 1980.
Kotilainen, Noora. "Puhdasta, suomalaista, nationalistista lihaa." In *Eläimet yhteiskunnassa*, edited by Elisa Aaltola and Sami Keto, 37–56. Helsinki: Into, 2015.
"Maailman Luonnon Säätiön Suomen Rahasto." *Suomen Luonto* 32 (1973): 106–7.
Mikola, Peitsa. "Neljännesvuosisata luonnonsuojelutyötä." *Suomen Luonto* 22 (1963): 75–7.
Nash, Linda. "The Agency of Nature or the Nature of Agency?" *Environmental History* 10 (2005): 67–9.
Peterle, Tony J. "DDT in Antarctic Snow." *Nature* 224 (1969): 620.
Pohja-Mykrä, Mari, Timo Vuorisalo, and Sakari Mykrä. "Organized Persecution of Birds of Prey in Finland: Historical and Population Biological Perspectives." *Ornis Fennica* 89 (2012): 1–19.
Räsänen, Tuomas. "Converging Environmental Knowledge: Re-evaluating the Birth of Modern Environmentalism in Finland." *Environment and History* 18 (2012): 159–81.
Räsänen, Tuomas. "Itämeren ympäristökriisi ja uuden merisuhteen synty Suomessa 1960-luvulta 1970-luvun puoliväliin." PhD diss., Turun yliopisto, 2015.
Ritvo, Harriet. *Noble Cows and Hybrid Zebras: Essays on Animals and History*. Charlottesville, VA: University of Virginia Press, 2010.
Russell, Edmund. *Evolutionary History: Uniting History and Biology to Understand Life on Earth*. New York: Cambridge University Press, 2011.
Salo, Miikka. "Uusiutuva energia ja energiajärjestelmän konsensus. Mekanistinen näkökulma liikenteen biopolttoaineiden ja syöttötariffin käyttöönottoon Suomessa." PhD diss., University of Jyväskylä, 2014.
Sayes, Edwin. "Actor-Network Theory and Methodology: Just What Does It Mean to Say That Nonhumans Have Agency." *Social Studies of Science* 44 (2014): 134–49.
Stjernberg, Torsten (zoologist, Finnish Museum of Natural History). In discussion with the author. December 2015.
Suominen, Teuvo. *Lintujemme katoava aateli*. Porvoo, Finland: WSOY, 1967.
Tatton, J. O'G., and J. H. A. Ruzicka. "Organochlorine Pesticides in Antarctica." *Nature* 215 (1967): 346–8.
Telkänranta, Helena. "Tiedotusvälineiden muuttuva maailmankuva." In *Laulujoutsenen perintö. Suomalaisen ympäristöliikkeen taival*, 222–5. Helsinki: Suomen Luonnonsuojeluliitto ry and WSOY, 2008.
"Vielä biosideista." *Suomen Luonto* 24 (1965): 32.
Wembridge, David. *Urban Mammals: A Concise Guide*. Stansted, UK: Whittet Books, 2012.

8 Cosmological Changes

Shifts in Human–Fish Relationships in Alaska's Bering Strait Region

Julie Raymond-Yakoubian and Vernae Angnaboogok

Introduction

This chapter addresses human–fish relationships amongst Alaskan Bering Strait Eskimo people,[1] and some ways those relationships have changed over time. Many species of fish[2] have long been critical subsistence resources for Bering Strait Eskimo people.[3] Fish are a valuable food source, provide important nutrition, are culturally preferred foods, and can be significant contributors to individual and community well-being and identity. One nineteenth-century explorer who traveled throughout the region and beyond even remarked that the Inupiat could be called "a nation of fishermen."[4]

As an illustration of the importance of fish to contemporary Bering Strait indigenous communities, consider the following statements made by region residents. One Teller resident noted that "[f]ish [is] very, very, very important … where my parents an' grandparents came from, which is upriver. Fish was central [to] life basically,"[5] and another stated the following: "If there was no tomcods, there'd probably be no Eskimos. Tomcods has pretty much saved a lot of Eskimos' lives, I guess."[6] Further, a Stebbins resident stated that "If you know how to hunt and fish, you're alright. You can live. You can survive. If you don't know how, you don't know nothing."[7] A Brevig Mission resident noted that "To us, [the fish are] the most important things we need to eat. The fish, we can't go without fish, ah?"[8] and an Elim resident noted that "I wouldn't know how to describe who I am, without having that subsistence of salmon."[9]

Despite their continued importance, fish are oft-ignored in discussions of highly marine mammal–dependent communities – such as many Bering Strait communities. While the relationship that many contemporary Bering Strait residents have with fish has undergone significant and disruptive changes over the last century, we think that analyzing human–fish relationships can provide important insights into Bering Strait Eskimo cosmology.

Alaska's Bering Strait region is located just south of the Arctic Circle, extending from the southern shores of Norton Sound to the northern Seward Peninsula, including the northern Bering Sea islands. The region is the homeland for three distinct cultural groups – the Inupiat, Yup'ik, and St. Lawrence Island Yupik people.

The information presented here about contemporary beliefs and practices is derived primarily from interviews and discussions with elders and active fishers, and from participant observation of fishing and other subsistence-related activities, carried out by the first author. The information comes from over 150 individuals in ten Alaskan Bering Strait communities.

There has been a shift in the main, dominant, standard beliefs and discourse about fish over time amongst the Indigenous People of this region from the early contact period, if not prior, to the present. These shifts have connections with broader cosmological shifts. We will discuss elements of early views about fish, contemporary views and major influences shaping these views, and examples of contemporary beliefs in practice, all of which show the shift in the dominant or standard beliefs and discourse over time as well as the complexities associated with this shift.

Earlier Beliefs about Fish

This chapter's focus is on a shift over time in indigenous views about fish and human–fish relationships in this region of Alaska. A precise moment cannot be pinpointed for this shift; rather, the shift has likely been uneven and occurred over time from what we know of earlier beliefs to how things are presently. A number of other terms might be applied to these earlier views – for example, "traditional," or "non-western." These terms would not be without problems, nor without merit either. It would be generally accurate to consider these earlier or older views to be what are often termed traditional and non-western, as long as one keeps in mind that tradition entails change, the field of social discourse evidences variety and heteroglossia,[10] hybridity occurs, and these terms and that to which they refer to are constant sources of debate and meta-discourse. Nonetheless, we will generally avoid using these terms for the sake of clarity and maintain focus on the shift over time, and will use the rather unartful but perhaps more accurate term "earlier" (or "older") views.

The following are likely key earlier Bering Strait Eskimo beliefs about fish, and thus what we take to be central elements of the earlier dominant or standard discourse about fish and human–fish relationships: (1) fish are persons, (2) fish are intelligent, (3) fish have agency, (4) fish have spiritual or other powers, (5) fish must be treated with respect, and (6) fish-persons have equivalence to human-persons.

Eskimo, Inuit and other indigenous groups from around the North share similar traditional beliefs about how animals are perceived and interacted with.[11] While there are few published ethnographic sources relating to earlier belief systems for the Bering Strait region, based on these sources,[12] it is reasonable to conclude that earlier Bering Strait Eskimo beliefs related to human–animal relationships were similar to those held by groups directly to their north and south (that is, in northwest and southwest Alaska), for whom we have more extensive ethnographic documentation on this matter.[13]

The earlier beliefs about fish noted above would have been in effect prior to the widespread introduction and later adoption of Christianity in the Bering Strait region (circa 1887),[14] and are similar to beliefs of other indigenous groups found across the North. These characteristics, which people associated with fish (and other animals and features of the environment), effected a particular kind of relationship which was, and in some cases still is, expressed through reciprocal relationships, based around ideas of respect. This relationship dictated human thought and behavior, and was possible because of the mutual intelligence, agency, and power of human-persons and non-human-persons.

Personhood is attributed to animals, including fish, in western Alaska Eskimo traditional belief systems. Agency is additionally an important characteristic of fish in these belief systems. Agency, as the term will be used in this chapter, can be attributed to a person if they have consciousness, intentionality, and decision-making ability. In the traditional belief system, fish possessed all the requirements necessary to be considered to have agency. While the above definition of agency has been developed fully here by the authors, our approach fits within a rich northern anthropological tradition addressing that and related concepts, where these elements of personhood and agency have been discussed extensively with regard to the ways many northern Indigenous People think about animals.[15]

For example, Fienup-Riordan has extensively documented the Yup'ik attribution of personhood and agency to fish and other animals. For instance, Chinook salmon could choose whose net to enter based on treatment they previously received from human fishers.[16] The remains of salmon were also never to be "left lying around or be thrown into the river," because if other fish saw them, "they would not continue upstream but return to the sea"; additionally, humans should avoid arguing about fish in season because fish hear discord amplified and become displeased.[17]

Contemporary Beliefs about Fish

We argue that Bering Strait Eskimo beliefs about the characteristics of fish, and proper or acceptable behavior towards fish, has significantly shifted since the introduction and widespread adoption of Christianity and other Western ideas and practices. The contemporary dominant or standard set of such beliefs can be outlined as such: (1) fish are not persons, (2) fish are sentient but not very intelligent, (3) fish have limited agency, (4) fish do not have spiritual or other powers, (5) fish must be treated with respect, and (6) fish are an economic and cultural resource. It is, however, important to remember that there is complexity in this field of belief and discourse, for example, as evidenced in variety and hybridity, including elements of earlier beliefs being seen.

For example, the majority of Bering Strait Eskimo residents interviewed do not think that fish and humans have any real equivalence or that fish have agency. Fish are considered by most to be sentient, but generally not very intelligent and to be operating mostly instinctually. Fish are also very much viewed as economic as well as cultural resources. Concurrently, people still feel that fish must be treated respectfully – the most common expression of which being dicta to not waste fish and to share fish with others.

The cultures and economies of Bering Strait communities have experienced many changes since earlier times. Wage labor, commercial goods, commodification of animal and fish resources, and other introductions have had various impacts on the region's indigenous residents. In terms of fish and human–fish relationships, three of the largest changes – all being related – are: (1) the Western rationalization of the local conceptualization of fish, (2) the commodification of fish (particularly through the introduction of commercial fishing), and (3) the Christianization of the conceptualization of fish and human–fish relationships.

What is meant here by the "Western rationalization" of the local conceptualization of fish is that the means by which fish operate as a logical entity in the local cosmology have shifted to the discursive grounds of Western thought. Several key modes in which this is accomplished are by metricalization, regulation, secularization, and biologizing. Fish and harvests are enumerated, tracked, permitted, and allocated; they are managed and regulated (especially in regard to addressing population "depletion"); and they are reduced to a finite set of Western scientific biological characteristics that amount to little more than behaviorally simple swimming matter. This new way of viewing fish fits well within the colonial culture's and economy's vision for Bering Strait region resource use, and in many ways is internalized in the everyday practices and perceptions of indigenous residents, reinforced in the many ways in which the simple act of harvesting a fish always, necessarily, and often traumatically involves not just fishers and fish but also the actors, such as wildlife enforcement staff, and texts, such as harvest regulations and reporting forms, of colonial management.

These processes have quite pervasive and deleterious effects. For example, metricalization entails not just a means to enumerate fish but also a technique by which fish is allocated inequitably to commercial over local subsistence fishers. Further, these processes indicate a powerful cosmological shift operating at a much deeper level. Sahlins' characterization of the juncture of Western ontology and epistemology in Judeo-Christian cosmology is most apt here: a "knowledge of natural things … reduced to sensory experience of the obdurate matter on which humanity was condemned to lay waste its powers."[18] This view of nature – human and otherwise – permeates Western fish and wildlife biology and management regimes. We therefore see, for example, the biological concern that more fishing or hunting results in less fish and animals, instead of the view from indigenous elders that "when you quit hunting animals or mammals, they'll just decrease in number. The more you hunt them they'll multiply."[19] We thus have the singularly Western view that, as Sahlins has cleverly put it, humans are basically beasts, rather than the view held in many cultures throughout human history (including, at least in earlier times, Bering Strait Eskimo cultures), that beasts are essentially human.[20]

This "Western rationalization" of fish is interconnected, and reinforced, by the second major force: the commodification of fish. Fish have always been an economic resource to Eskimo communities, whether through trade or through sale on a small scale. Commercial fishing, on the scale it is seen today, however, is a relatively new activity for indigenous Bering Strait residents,[21] and one which has taken on substantial importance for many families. Commercial fishing has become an integral part of the yearly cycle of resource harvesting, and income from fishing can facilitate participation in subsistence activities. Participation in a market economy surrounding fish – an economy that extends far beyond the local and into the global – has led to a necessary adoption, reinforcement, and incorporation of Western concepts about what kind of things fish are and can be, and how human beings relate to them – all as part of the necessary process of trying to operate within the current colonial regime.

The third and final key force we consider is an explicit Christianization of conceptualizations of fish and human–fish relationships. Whereas what we have characterized as the earlier views of fish are marked by, among other things, an animistic quality, there has been a shift in that many now understand fish as being either (Christian) God-directed or motivated through what might be called a "Christian animatism," that is, fish as part of God's blessing on earth. We will now turn to how some of the above is manifested in everyday life.

Beliefs in Practice

As noted, one key cosmological shift has been from a more animistic worldview – where all things were part of a broader interconnected system and the various entities contained within (for example, people and fish) had spirits and personhood – towards one which revolves more around a Christian God. This also entails a concomitant loss of particular kinds of agency in conceptualizations of fish, and extends not only to ideas about fish but also ideas about human–fish relationships. This Christian God–centered view sometimes manifests as what could be called a Christian animatism via the notion of "God's blessing." For example, one individual noted that

> Salmon is a blessing from God, because God created everything. And He's our creator. He created the salmon. It's a blessing from God because he gave it to be part of our food. God created the salmon to be part of our way of life.[22]

Another interviewee noted, "God gave us these fish, so you got to take care of 'em, not play with them. Because some day you might starve to death or whatever, if there's no fish. Just don't play with what God gave you – fish,"[23] and another stated that "People say I'm lucky to catch fish. I'm very lucky to catch fish. Yeah, lucky. Yes, but who gave it to us? God did."[24]

The Christian God–centered view also sometimes manifests as a more "God-directed" notion. Many individuals expressed a belief that it is not a fish that makes a conscious decision or choice to come to your net or hook, but rather that God is directing the fish to you, or you to the fish. For example, an Elim fisher noted that "He [God] just directs us to [the fish]. He directs us to know where it's at."[25]

Sometimes what is seen with these cosmological shifts is that many contemporary practices have surficially remained the same as older ones, but deeper engagements with people often reveal that those practices' underpinnings differ from what we know of older ones. In earlier times, an important reason why people shared their food widely and did not waste food was because of their close personal, reciprocal relationships with animals. Both practices – sharing and not wasting – were pleasing to animals and were two of many acts that would help ensure the return of animals for harvest in future seasons. For example, a harvested animal would, in death, still be aware of how it was being treated by humans; if treated properly, for example, shared, this would please the animal and it would be more likely to return the next year to offer itself to humans for harvest. Rather than trying to please the animal, a view many hold today is that sharing *pleases God*, and that the relationship is now between humans and God rather than between humans and animals. For example, one Shishmaref elder expressed some of his views on the agency of fish and why a fisher might be successful:

> I always think that it's mostly God. You know … God knows, the Heavenly Father above. If you're not stingy or anything, the more we give, the more you get back. In some way. And that would make you feel good too, you know – to give.[26]

Another elder from Shishmaref stated:

> The old saying is, long ago, is that the more you give away, the more you'll get back. An' sometimes, they're always right. If you give away fish, the next day your net for some reason just happens to be a little bit fuller. I didn't know if somebody's watching us, but it seems to be true that the more food you give away, it'll always come back. So we aren't afraid to give away food.[27]

This elder implies that there is some kind of higher power involved, and many region residents now specifically assign that power to a Christian God, such as is described by this Elim resident:

> I would not consider it the animal or a spirit. I would consider it God watching out for his people. If He sees our actions, He sees what's in our heart and He sees if we give grudgingly. But, if we share, or we get our catch and we take care of it …, you get blessed for giving, and for caring for your own. But I would say it's God, it's not the spirit of the animal.[28]

In order to participate in contemporary subsistence and small-scale local commercial harvests, Bering Strait residents necessarily become enmeshed in processes of metricalization. This is, indeed, the grounds of discourse with managers and enforcement actors. One result of this is a shift in discourse at the local level. People must keep tallies of harvests for managers, debate the merits of escapement counts and counting technologies, engage in policy debates over allowable fish bycatch, and so on. While this does manifest in some ways with the adoption of these new grounds for thinking and talking about fish and human–fish relationships, it also offers examples of the complexity of relationships between "tradition" and "modernity," and between different ways of thinking, knowing, and talking. As elsewhere, of course, we see hybridization of "traditional" and "non-traditional" positions at conscious (for example, pushback against the importance of metricalization) and unconscious levels. We must also not assume that participation in this process is simple agreement. Local residents are in many cases forced by the indecency of others[29] into situations where certain values are pitted against each other, for example sharing versus protecting resources, regulation versus local management. Given the importance of these fish-related activities to nutrition, culture, identity, and personal economics, participation in these processes of metricalization can hardly be seen as a choice, and is one of the strongest examples of the ways raw power enters into cosmological interfaces.

In earlier times, all animals, including fish, were considered to have agency, which included consciousness, intelligence, and decision-making ability. Many Bering Strait residents today, however, view fish as sentient but not very intelligent. Fish are seen as operating primarily on instinct, and are sentient in that they are able to find their natal rivers and avoid visible obstacles, but are no longer seen as having complex intelligence. For example, when asked if fish have an awareness of people and their actions, a Wales resident replied:

> I think so … Yes, I do.… Because you have to keep your net clean, like I said. Fish can tell when their cousins are sitting there an' not moving. Then they know.… They'll swim, long as they don't see it. But if they see it, they see lot of their cousins hanging there, [they say] "I don't think I wanna swim that way."[30]

A Shishmaref interviewee noted:

> I think we're the smart ones, and they're [fish] still developing. … We keep improving on our lifestyle. Learn to do a lot more things. And they're still fish, they still do the same thing. But they get smart too, if there's too much disruption someplace they move or die off.[31]

The individuals above, like others, have a view of fish that allows for a degree of intelligence. Their avoidance, however, is not premised on higher level intelligence that would include information such as *whose* net they were avoiding, or for what particular reason they were avoiding it, for example. Interestingly, some other animals are viewed as being highly intelligent. When asked if he thinks fish are smart, another Shishmaref elder replied:

> Like humans, some are smarter than others. … The [animals] with experience, other than fish, like seals, they've been hunted before an' they know the habits of the hunter. An' they're very, very smart.[32]

One interviewee from Stebbins stated that:

> They can be smart, they evade invaders. When they're being chased by another predator, their first instinct is … take off. Go to a sheltered area. So that's one way they could be smart. If they weren't intelligent in their own way, they wouldn't survive, right?[33]

And another Stebbins resident remarked:

> I'm glad they ain't too smart [laughs]. If they [fish] were too smart, the sea mammals would have a hard time getting their daily food – if they were smart. They have a purpose, you know, so that's where they are.[34]

On the whole, for indigenous region residents, fish are, by and large, seen as being not very intelligent. A diminished sense of fish intelligence compared to earlier views is one key characteristic of the contemporary main, standard set of beliefs and discourse on fish. The examples immediately above express beliefs that many indigenous region residents hold – fish are able to recognize some human activities (for example, net fishing) as something to be avoided, but they lack a higher-order, complex intelligence – one that might be comparable to human intelligence – some other animals are thought to possess. It is possible there is a relationship in this to the fact that fish are the most commodified of all the animals in the Bering Strait Eskimo environment, and those animals that are largely not commodified are still often portrayed as more intelligent, as giving themselves to hunters, that is, with more traditional views. One might argue that fish are also some of the least "individualistic" animals, and given the high value placed in Western cosmology on individualism for rationality, this may be connected with the better intellectual station more solitary animals, such as moose or beluga whales, are accorded compared to fish. The ease of harvest may also be connected to this differential; fish are largely seen as easy to catch, whereas harvesting other animals takes more effort. There is (even evident in some examples above) some variation within such views, as well as views that diverge from this standard. These illustrate variety and hybridity at the social and individual level regarding various discourses and beliefs such as the co-existence or blending of old and new, and indigenous and non-indigenous, discourses and beliefs. Nonetheless, a shift in the main, standard discourse and set of beliefs is clearly evident.

The importance of commodification to shifting beliefs is likely quite important. For example, a commonly held older belief noted above is that "when you quit hunting animals or mammals, they'll just decrease in number. The more you hunt them they'll multiply."[35] Another region elder who held many older beliefs (for example, regarding proper ritual treatment of harvested seals, the function of sharing, and of animals giving themselves to hunters) noted that, "[a]s far as I know we have a pretty healthy run for [that river], although it is a small run, it sustains itself even though we do harvest a few by seining or rod and reel."[36] The "even though" in the preceding quote is notable in that it indicates a divergence away from the view that harvesting results in population increases towards the more Western notion that harvesting causes population depletion. This particular interviewee was heavily involved in commercial fishing, and was particularly concerned about the impact of commercial fishing on salmon populations due to personal experience. It is hypothesized here that this encounter with this Western mode of economic action pushed a shift at the individual level with this interviewee away from traditional views to Westernized ones, while other traditional views regarding other subsistence resources and activities were maintained.

Another example of a different type of hybridization and complex mixing of beliefs surrounding fish and human–fish relationships pertains to subsistence versus small-scale commercial fish harvests. These activities are regulated differently in Alaska, and the former is understood as being a traditional activity for providing food for people, families, sharing, and bartering, while the latter is seen as means for providing income – in many cases, income which goes towards costs of conducting subsistence activities. In practice, however, there is often very little from a "technical" or "mechanical" perspective which separates the two activities, which in many cases are even done at the same time.[37] Yet, it is a common sentiment in the region that selling subsistence foods is a bad thing, while selling small-scale commercial harvests is acceptable. In these cases, we can see here how the fish itself becomes ontologically different (for example, sacred versus profane) depending on the mindset and intentionality of the fisher at the exact moment of its harvest. That this is possible is most likely a result of the broader shifts which we have discussed in this chapter having occurred.

The above examples demonstrate some of the complex ways in which broader cosmological constructs manifest themselves in terms of various aspects of conceptualizations of fish and human–fish relationships. These manifestations play out in everyday life, speaking to extremely locally important processes of history, politics, and identity felt at both the individual and social levels.

Concluding Remarks

Much of the above has focused on the overall shift in the dominant, or standard, discourse and set of beliefs about fish. We have additionally stressed the also-evident variety and complexity related to this domain as well, for example, in the field's heteroglossic nature and in instances of hybridity. In these final comments, we would like to highlight the historical complexities and profound interconnections with broader sociopolitical realities associated with these discourses and beliefs, and their shifts.

As we have intimated, there are complex relationships, interweaving discourses, and hybridized realities about, between, and emanating from practices, beliefs, meanings, history, and values which are involved in the actual lived reality of the human–fish relationships discussed in this chapter and the cosmologies in which they are situated. While space has constrained us to broad outlines, our contribution should not be taken to imply that history is writ from outside, even if the dominant views today are strongly influenced by forces historically external to indigenous communities. The authors feel that history's reality and meaning must be understood through local generative processes. History, we feel, is not writ from the outside, and we concur with Sahlins in seeking to understand modernity as indigenized, as well as with his view that the confrontation between the West and the Rest in colonial history is best seen "as a triadic historical field, including a complicated intercultural zone where the cultural differences are worked through in political and economic practice."[38]

How people relate to animals and the environment is critical to food security, physical and mental health, and the formation and maintenance of cultural identity. There have been many changes to Bering Strait Eskimo life that have resulted in a variety of new forms of belief and practice regarding fish which, in addition to their historical contexts, need to be acknowledged in order to have a fuller understanding of contemporary indigenous life. The attendant shifts in cosmology and their complex manifestations in a generally under-examined aspect of region life (views on fish and human–fish relationships) have great importance for the ways in which Bering Strait people interact with animals and the environment and engage with – and are engaged by – the wider world.

Acknowledgments

The research this chapter is based on was funded in part by Kawerak, the U.S. Fish and Wildlife Service's Office of Subsistence Management (Study 10-151), the Arctic-Yukon-Kuskokwim Sustainable Salmon Initiative (projects 601, 712 and 1333), Alaska EPSCoR NSF award #OIA-1208927 and the State of Alaska, the North Pacific Research Board, and the Norton Sound Economic Development Corporation. The Tribal Councils in Brevig Mission, Diomede, Golovin, Koyuk, Saint Michael, Stebbins, Teller, Wales, White Mountain, and, particularly, Elim collaborated on the initial research, which could not have been conducted without the generosity of those communities' tribal members. We also thank Brenden Raymond-Yakoubian, Mary David, and the participants in the 2014 Urban Rural Wilderness workshop in Turku for helpful comments on chapter drafts.

Notes

1 In Alaska's Bering Strait region, many Indigenous Peoples and organizations use the term "Eskimo" (or more specific designations such as "Inupiat," "Yup'ik," or "St. Lawrence Yupik") as opposed to "Inuit" for self-identification. We acknowledge the term is not widely used outside Alaska.

2 This chapter focuses on fish harvested by Bering Strait Indigenous Peoples, i.e., the five Pacific salmon species as well as non-salmon fish, for example, tomcod, herring, and smelt.

3 Raymond-Yakoubian, *Climate-Ocean Effects on Chinook Salmon*; Raymond-Yakoubian and Raymond-Yakoubian, "*Always Taught Not to Waste*"; Raymond-Yakoubian "*When the Fish Come*"; Burch, *Social Life in Northwest Alaska*; Ray, "Nineteenth Century Settlement"; Nelson, "The Eskimo about Bering Strait."

4 Beechey, *Narrative of a Voyage*.

5 Raymond-Yakoubian, "*When the Fish Come*," 83.

6 Ibid.
7 Ibid., 105.
8 Topkok and Raymond-Yakoubian, *Food for the Soul*, 47.
9 Elim resident interviewed by J. Raymond-Yakoubian on May 17, 2011.
10 See, for example, Bakhtin, *The Dialogic Imagination* and also Sahlins, *Waiting for Foucault, Still.*
11 See, for example, Berkes, *Sacred Ecology*; Brightman, *Grateful Prey*; Hallowell, "Ojibwa Ontology"; Nadasdy, *Hunters and Bureaucrats*; Nelson, *Make Prayers to the Raven*; Tanner, *Bringing Home Animals*; Wenzel, *Animal Rights, Human Rights*; Willerslev, *Soul Hunters*.
12 For example, Fitzhugh, Kaplan, and Collins, *Inua*; Kingston, *Traditional Ecological Knowledge*; Michael, *Lieutenant Zagoskin's Travels*; Nelson, *The Eskimo about Bering Strait*; Oquilluk, *People of Kauwerak*; Ray, "Nineteenth Century Settlement"; Ray, *Ethnohistory in the Arctic*; Ray, *The Eskimos of Bering Strait*; Raymond-Yakoubian, "*When the Fish Come*"; Raymond-Yakoubian and Raymond-Yakoubian, "*Always Taught Not to Waste*."
13 E.g., Burch, *The Eskimos*; Burch, *The Iñupiaq Eskimos*; Bodenhorn, "The Animals Come to Me"; Fienup-Riordan, *The Nelson Island Eskimo*; Fienup-Riordan, *When Our Bad Season Comes*; Fienup-Riordan, *Eskimo Essays*; Fienup-Riordan, *Boundaries and Passages*; Fienup-Riordan, "Yaqulget Qaillun Pilartat"; Fienup-Riordan, *Hunting Tradition*; Hensel, *Telling Our Selves*; Kawagley, *A Yupiaq Worldview*; Lantis, *Alaskan Eskimo Ceremonialism*; Nelson, *Hunters of the Northern Ice*; Spencer, *The North Alaskan Eskimo*.
14 Almquist, *Covenant Missions in Alaska*; Burch, "The Inupiat and the Christianization of Arctic Alaska"; Olsson, *By One Spirit*; Savok, *Jesus and the Eskimo*; VanStone, "Alaska Natives."
15 E.g. Berkes, *Sacred Ecology*; Brightman, *Grateful Prey*; Fienup-Riordan, *Boundaries and Passages*; Fienup-Riordan, *Hunting Tradition*; Nadasdy, *Hunters and Bureaucrats*; Nelson, *Make Prayers to the Raven*; Tanner, *Bringing Home Animals*; Wenzel, *Animal Rights, Human Rights*; Willerslev, *Soul Hunters*.
16 Fienup-Riordan, *Hunting Tradition*, 52.
17 Fienup-Riordan, *Boundaries and Passages*, 120.
18 Sahlins, "The Sadness of Sweetness," 411.
19 Brevig Mission resident interviewed by J. Raymond-Yakoubian, May 21, 2008.
20 Sahlins, *The Western Illusion*, 2. To put it more explicitly, Sahlins here counterposes a particularly Western (for example, Hobbesian) view that humans are seen as having a primordial unruly "animalistic" inner nature to overcome with the view in many other cultures that animals are basically human in their nature.
21 Commercial fisheries with a reasonable semblance to what is seen today in terms of their regulation and structure have been in place in the region since the early 1960s.
22 Elim resident interviewed by J. Raymond-Yakoubian, May 31, 2011.
23 Elim resident interviewed by J. Raymond-Yakoubian, July 2, 2011.
24 Elim resident interviewed by J. Raymond-Yakoubian, June 12, 2011.
25 Elim resident interviewed by J. Raymond-Yakoubian, June 15, 2011.
26 Raymond-Yakoubian, "*When the Fish Come*," 135.
27 Shishmaref resident interviewed by J. Raymond-Yakoubian, January 31, 2012.
28 Elim resident interviewed by J. Raymond-Yakoubian, June 12, 2011.
29 For example, large-scale commercial interests removing enormous amounts of fish from the oceans, politicians who do not recognize the status of Indigenous Peoples, scientists who do not recognize the value of traditional knowledge, and resource managers who inequitably enforce regulations on indigenous communities while permitting gross violations by non-indigenous people. Raymond-Yakoubian, "Participation and Resistance;" Raymond-Yakoubian, Raymond-Yakoubian, and Moncrief, "The Incorporation of Traditional Knowledge into Alaska Federal Fisheries Management."

30 Wales resident interviewed by J. Raymond-Yakoubian, August 30, 2011.
31 Shishmaref resident interviewed by J. Raymond-Yakoubian, January 31, 2012.
32 Shishmaref resident interviewed by J. Raymond-Yakoubian, October 11, 2011.
33 Stebbins resident interviewed by J. Raymond-Yakoubian, January 26, 2012.
34 Stebbins resident interviewed by J. Raymond-Yakoubian, January 26, 2012.
35 Brevig Mission resident interviewed by J. Raymond-Yakoubian, May 21, 2008.
36 Golovin resident interviewed by J. Raymond-Yakoubian, April 18, 2009.
37 See also Reedy-Maschner, "Entangled Livelihoods" and Raymond-Yakoubian and Raymond-Yakoubian, "*Always Taught Not to Waste*."
38 Sahlins, "Goodbye to *Tristes Tropes*," 486, 495.

Bibliography

Ethnographic Material

All audio and transcripts of the interviews are in the possession of the author.

Literature

Almquist, Arden. *Covenant Missions in Alaska*. Chicago: Covenant Press, 1962.

Bakhtin, Mikhail. *The Dialogic Imagination*. Edited by Michael Holquist, translated by Caryl Emerson and Michael Holquist. Austin, TX: University of Texas Press, 1982.

Beechey, Frederick William. *Narrative of a Voyage to the Pacific and Beering's Strait, to Co-operate with the Polar Expeditions Performed in his Majesty's Ship Blossom, under the Command of Captain F. W. Beechey, R. N. in the Years 1825, 26, 27, 28*. Vol. 2. London: Colburn and Bentley, 1831.

Berkes, Fikret. *Sacred Ecology*. New York: Routledge, 1999.

Bodenhorn, Barbara. "The Animals Come to Me, They Know I Share: Iñupiaq Kinship, Changing Economic Relations and Enduring World Views on Alaska's North Slope." PhD diss., University of Cambridge, 1989.

Brightman, Robert. *Grateful Prey: Rock Cree Human-Animal Relationships*. Saskatchewan: Canadian Plains Research Center, 1973.

Burch, Ernest, Jr. *The Eskimos*. Norman, OK: University of Oklahoma Press, 1988.

Burch, Ernest, Jr. "The Inupiat and the Christianization of Arctic Alaska." *Etudes/Inuit/Studies* 18 (1994): 81–108.

Burch, Ernest, Jr. *The Iñupiaq Eskimos of Northwest Alaska*. Fairbanks, AK: University of Alaska Press, 1998.

Burch, Ernest, Jr. *Social Life in Northwest Alaska: The Structure of Iñupiaq Eskimo Nations*. Fairbanks, AK: University of Alaska Press, 2006.

Fienup-Riordan, Ann. *The Nelson Island Eskimo: Social Structure and Ritual Distribution*. Anchorage: Alaska Pacific University Press, 1983.

Fienup-Riordan, Ann. *When Our Bad Season Comes: A Cultural Account of Subsistence Harvesting and Harvest Disruption on the Yukon Delta*. Anchorage: Alaska Anthropological Association, 1986.

Fienup-Riordan, Ann. *Eskimo Essays: Yup'ik Lives and How We See Them*. Chapel Hill, NC: Rutgers University Press, 1990.

Fienup-Riordan, Ann. *Boundaries and Passages: Rule and Ritual in Yup'ik Eskimo Oral Tradition*. Norman, OK: University of Oklahoma Press, 1994.

Fienup-Riordan, Ann. "'Yaqulget Qaillun Pilartat (What the Birds Do)': Yup'ik Eskimo Understandings of Geese and Those Who Study Them." *Arctic* 52 (1999): 1–22.

Fienup-Riordan, Ann. *Hunting Tradition in a Changing World: Yup'ik Lives in Alaska Today*. London: Rutgers University Press, 2000.

Fitzhugh, William, Susan Kaplan, and Henry Bascom Collins. *Inua: Spirit World of the Bering Sea Eskimo*. Washington, DC: Smithsonian Institution Press, 1982.

Hallowell, Alfred Irving. "Ojibwa Ontology, Behavior, and World View." In *Culture in History: Essays in Honor of Paul Radin*, edited by Stanley Diamond, 19–52. New York: Columbia University Press, 1960.

Hensel, Chase. *Telling Our Selves: Ethnicity and Discourse in Southwestern Alaska*. New York: Oxford University Press, 1996.

Kawagley, Angayuqaq Oscar. *A Yupiaq Worldview: A Pathway to Ecology and Spirit*. Prospect Heights: Waveland Press, 1995.

Kingston, Deanna. *Traditional Ecological Knowledge of Ugiuvangmiut (King Island) Iñupiaq Walrus Hunters*. Report for the Pacific Walrus Conservation Fund, 2005.

Lantis, Margaret. *Alaskan Eskimo Ceremonialism*. New York: Augustin, 1947.

Michael, Henry, ed. *Lieutenant Zagoskin's Travels in Russian America, 1842–1844: The First Ethnographic and Geographic Investigations in the Yukon and Kuskokwim Valleys of Alaska*. Toronto: University of Toronto Press, 1967.

Nadasdy, Paul. *Hunters and Bureaucrats: Power, Knowledge, and Aboriginal-State Relations in the Southwest Yukon*. Vancouver: University of British Columbia Press, 2003.

Nelson, Edward William. "The Eskimo about Bering Strait." In *Eighteenth Annual Report of the Bureau of American Ethnology, 1896–1897*, part 1, 3–518. Washington, DC: Government Printing Office, 1899.

Nelson, Richard. *Hunters of the Northern Ice*. Chicago: University of Chicago Press, 1969.

Nelson, Richard. *Make Prayers to the Raven: A Koyukon View of the Northern Forest*. Chicago: University of Chicago Press, 1983.

Olsson, Karl. *By One Spirit*. Chicago: Covenant Press, 1962.

Oquilluk, William A. *People of Kauwerak: Legends of the Northern Eskimo*. With the assistance of Laurel L. Bland. Anchorage: Alaska Methodist University, 1973.

Ray, Dorothy Jean. "Nineteenth Century Settlement and Subsistence Patterns in Bering Strait." *Arctic Anthropology* 2 (1964): 61–94.

Ray, Dorothy Jean. *Ethnohistory in the Arctic: The Bering Strait Eskimo*. Ontario: Limestone Press, 1983.

Ray, Dorothy Jean. *The Eskimos of Bering Strait, 1650–1898*. Seattle: University of Washington Press, 1992.

Raymond-Yakoubian, Brenden and Julie Raymond-Yakoubian. *"Always Taught Not to Waste": Traditional Knowledge and Norton Sound/Bering Strait Salmon Populations*. Arctic-Yukon-Kuskokwim Sustainable Salmon Initiative Project 1333. Nome, AK: Kawerak, 2015.

Raymond-Yakoubian, Julie, Brenden Raymond-Yakoubian and Catherine Moncrief. "The Incorporation of Traditional Knowledge into Alaska Federal Fisheries Management." *Marine Policy* 78 (2017): 132–142.

Raymond-Yakoubian, Julie. *Climate-Ocean Effects on Chinook Salmon: Local Traditional Knowledge Component*. Arctic-Yukon-Kuskokwim Sustainable Salmon Initiative Project 712. Nome, AK: Kawerak, 2009.

Raymond-Yakoubian, Julie. "Participation and Resistance: Tribal Involvement in Bering Sea Fisheries Management and Policy." In C. Carothers, K.R. Criddle, C.P. Chambers, P.J. Cullenberg, J.A. Fall, A.H. Himes-Cornell, J.P. Johnsen, N.S. Kimball, C.R. Menzies, and E.S. Springer (eds.), *Fishing People of the North: Cultures, Economies, and Management Responding to Change*. Alaska Sea Grant, University of Alaska Fairbanks, 2012.

Raymond-Yakoubian, Julie. *"When the Fish Come, We Go Fishing": Local Ecological Knowledge of Non-Salmon Fish Used for Subsistence in the Bering Strait Region*. USFWS, Office of Subsistence Management, Study No. 10-151. Nome, AK: Kawerak, 2013.

Reedy-Maschner, Katherine. "Entangled Livelihoods: Economic Integration and Diversity in the Western Arctic." *Alaska Journal of Anthropology* 7 (2009): 135–146.

Sahlins, Marshall. "The Sadness of Sweetness: The Native Anthropology of Western Cosmology." *Current Anthropology* 37 (1996): 395–415.

Sahlins, Marshall. "Goodbye to *Tristes Tropes*: Ethnography in the Context of Modern World History." In *Culture in Practice: Selected Essays* by Marshall Sahlins, 471–500. New York: Zone Books (1993) 2000.

Sahlins, Marshall. *Waiting for Foucault, Still*. Chicago: Prickly Paradigm Press, 2002.

Sahlins, Marshall. *The Western Illusion of Human Nature*. Chicago: Prickly Paradigm Press, 2008.

Savok, Fred. *Jesus and the Eskimo*. Fairbanks, AK: HLC Publishing, 2004.

Spencer, Robert. *The North Alaskan Eskimo: A Study in Ecology and Society*. Washington, DC: Government Printing Office, 1959.

Tanner, Adrian. *Bringing Home Animals: Religious Ideology and Mode of Production of the Mistassini Cree Hunters*. New York: St. Martin's Press, 1979.

Topkok, Meghan, and Julie Raymond-Yakoubian, compilers. *Food for the Soul: Bering Strait Non-Salmon Fish Preparation and Recipes*. Nome, AK: Kawerak Inc., 2013.

VanStone, James. "Alaska Natives and the White Man's Religion: A Cultural Interface in Historical Perspective." In *Exploration in Alaska, Captain Cook Commemorative Lectures*, edited by Antoinette Shalkrop, 175–179. Anchorage: Cook Inlet Historical Society, 1980.

Wenzel, George. *Animal Rights, Human Rights: Ecology, Economy and Ideology in the Canadian Arctic*. Toronto: University of Toronto Press, 1991.

Willerslev, Raine. *Soul Hunters: Hunting, Animism, and Personhood among Siberian Yukaghirs*. Los Angeles: University of California Press, 2007.

9 The Baltic Herring as Agents in the Socio-Ecological System in Rymättylä Fisheries

Kirsi Sonck-Rautio

Introduction

In this article, I examine the agency of the Baltic Herring (*Clupea harengus membras*) in regards to the Rymättylä socio-ecological system (SES), the local ecological knowledge (LEK) possessed by the fishermen in this particular system, and how they are related to adapting to global changes. I investigate the notion of animal agency and examine important concepts in this particular context after which I focus my attention on the former winter seining community of Rymättylä.

Rymättylä is a former municipality[1] in the Archipelago of southwest Finland, surrounded by the Archipelago Sea, which is part of the Baltic Sea. In the Nordic hemisphere, adaptation to freezing temperatures and an access to limited resources over winter used to be essential for survival. Rymättylä was a community where subsistence during the winter months relied on the traditional winter seining livelihood, until its seemingly rapid decline from the 1960s onwards. By the end of the 1990s, the winter seining livelihood had ceased to exist and the community had adapted to the changed circumstances.[2] The case of Rymättylä is a good example of a community that was once formed around a nature-based subsistence, but which has been influenced by many global factors and changed as a consequence.

This case can be used to analyze the adaptation process by seeing the community and the environment surrounding it as an SES.[3] I briefly go through the meaning of an SES analysis in the field of adaptation to global changes research and fisheries management research, and examine the importance of adding animal agency as a factor to the system. I analyze the ways the agency of the herring in the Rymättylä SES can be examined from the ethnoichthyological point of view, and why it would be important to do so. My objective here is to examine and analyze the ways in which animal agency and LEK concerning this agency could be applied in order to create a more profound understanding of the local sustainabilities, and how this understanding could help in planning fisheries management and adaptation strategies.

Animal Agency

Sarah McFarland and Ryan Hediger have pointed out that one type of agency could be expressed as the need for liberty, the right to life, and the pursuit of happiness. Quite often we human beings ignore these needs in regard to animals, because they seem to lack those specific cognitive abilities that separate them from human beings.[4] The agency of animals is indeed quite often denounced based on the arguments that they are not being conscious, decision-making, moral beings, and they lack cognitive abilities. The nature of the species certainly varies, and there are many unanswered questions regarding the general determinants of agency, especially when it concerns a gregarious and shoaling species such as the Baltic Herring.

If we consider decision-making to be one of the attributes of agency, we need to consider the ability of fish to make decisions. Experts in collective behaviour, David Sumpter et al. showed that a small group of sticklebacks (*Gasterosteus aculeatus*) reach consensus when they are forced to choose which fish to follow between two different replicas. The conclusion was that groups of fish do make a consensus-based decision and they do so because it is beneficial to them.[5]

One mystery closely connected to the agency of shoaling fish is the question of how the fish shoal determines its movements. If it is totally random, there is really no room for conscious decision-making. However, biologist Stephan G. Reebs studied gregarious Golden Shiners (*Notemigonus crysoleucas*), and found that in the case of regular migration, most of the shoaling fish are almost certainly aware of the destination to which they are heading. This can be explained by young fish learning from the more experienced individuals.[6] There might even be some leaders that always lead the shoal. The leaders are, however, replaceable, and the shoal will find its way even if the leaders are removed.[7] Therefore, one could conclude that in general, shoaling fish are capable of not only of making decision but conscious decision-making.

One argument against animal agency has been the question of the cognitive ability of animals. Biologist Iain D. Couzin examined collective cognition in animal groups, and pointed out that for a long time there has been an assumption that the decision-making among group animals was either somewhat telepathic or a synchronized response to a certain individual's commands. Recent studies, however, show that collective decision-making mechanisms among animal groups, such as flocking birds and shoaling fish, share functional characteristics. To go beyond theoretical debate, scientists have developed computer simulations based on mathematical modeling techniques, in order to study collective behavior. What was found was that "collective behaviour can arise from repeated and local interactions and need not be explicitly coded as a global blueprint of template."[8] In other words, it seems that fish are able to observe, learn, and adapt.

Fish are genetically individuals, but living in groups can increase their capacity to observe and react to their environment. Interactions with other group members actually enable individual fish to bypass their cognitive restraints, and gain access to context-dependent information and to information that is spatially and temporally integrated. From this type of interaction, the outcome can be more accurate decision-making even in the case of distractions and hesitation.[9] With these arguments in mind, it would not be reliable at all to claim that group animals, such as shoaling fish, do not have cognitive abilities.

In addition, a recent study of Shaun S. Killen et al. has suggested that there are certain types of individual fish that are less likely to be caught by trawlers. It is therefore very possible that harvest-induced selection and natural selection support each other with regard to their influence on the future type of fish stock.[10] Therefore, it is also the individual traits of fish that influence the SES in general.

What Is "Acting"?

Even though I have concluded that shoaling and schooling fish possess characteristics that can be considered to be the "determinants" of animal agency, it is still relevant to question whether these determinants are those required, or should agency be determined by other attributes? Anthropologist and philosopher Bruno Latour believes this to be the case, and he problematizes the meaning and definition of the term Actor, since it implies that there is a person who comes onto the stage. However, as Latour points out, it is never clear "who and what is acting since we are never alone in acting."[11] Latour also states that agencies are always present in "making some difference to the state of affairs, transforming some As into Bs through trials with Cs."[12] Therefore agency, instead of being a *who*, is actually a *what*. To put it simply, anything having an effect has agency. With this notion, I argue it is profitable and enlightening to view animals as significant actors when examining communities, social systems, and adaptation.

Socio-Ecological Systems and Fisheries Management

Adapting to changing global circumstances has become inevitable for many local communities. Since the crisis in marine environments, global warming, and overexploitation of fish stocks are crucial global problems, many actors are trying to develop the most sustainable policy-making solutions for the management of fisheries.[13] One of the ways to approach this issue is by ecosystem modeling, which includes numerous factors such as diet, consumption, fishing, prey-predator relationships, and change in biomass.[14]

By analyzing multiple research studies conducted in small-scale fisheries, it has been concluded that the most beneficial way of developing successful marine management strategies is to facilitate a local SES.[15] SES research sees local communities as entities that consist of social and ecological subsystems that are mutually interactive.[16] By analyzing the interconnections and relationships between these entities, researchers can find solutions for sustainable adaptation. In a 2001 IPCC assessment, geographer Barry Smit and climatologist Olga Pilifosova concluded that adaption is dependent on the adaptive capacity of a system influenced by stimuli and that "the adaptive capacity of communities is determined by their socioeconomic characteristics."[17] An SES is often applied when analyzing these relevant socioeconomic characteristics.

The applicability of the SES analysis is therefore rather extensive. Anthropologists and ethnologists have recently become more engaged in conservation and community-level adaptation research, and local socio-cultural systems are receiving more of their much-deserved attention.[18] However, the perspective of animal agency is missing as regards the functioning of the SES. The dimension of animal agency is still highly under-represented in SES analysis, and by engaging the perspectives of ethnoichthyology and LEK it is possible to find more economically, socially, ecologically, and culturally sustainable solutions to problems concerning conservation and adaptation to global changes.

Fish Agency and Local Ecological Knowledge

As shown above, there are ways to examine the agency of fish biologically, but if we decide to follow the concept of agency as defined by Latour, that is, as what, instead of who, then one of the best ways to approach the subject in the context of fisheries is to study the LEK of the fishermen.

Anthropologist Julie Cruikshank defined LEK as the tacit knowledge that is acquired by conducting extensive observations of an area and reproduced in everyday life with behavior and speech.[19] This type of knowledge is transmitted via oral history or shared collectively among the users of this particular resource.[20]

Applying LEK in search of a better means of management is not a new notion, and conservation research involving LEK among local fishermen has been conducted, for example, in Alaska, Australia, and Sweden.[21] However, when examining the agency of the animals, which are important to the local coastal SES, I find LEK and ethnoichthyology to be enlightening points of views.

The Story of the Baltic Herring and Rymättylä

The Baltic Herring (Figure 9.1) is the subspecies of the Atlantic Herring (*Clupea harengus harengus*), and it is well adapted to the brackish water of the Baltic Sea. The size of the average Baltic Herring is 14–18 cm and 30–90 grams. The Baltic Herring is a gregarious species and they migrate.[22] The life cycle of the herring consists of their birth, migration, foraging, and reproduction, that is, spawning. Each of these actions contained features that were crucial for the Rymättylä winter seining SES to function.

Figure 9.1 The Baltic Herring.

Source: Painting by Riinaliina Eräpohja.

Biologists Marjut Rajasilta and researchers at the Seili Archipelago Research Institute have studied the Baltic Herring for 30 years, and found that the inner parts of the Archipelago Sea are popular among the herring for reproduction.[23] The herring prefer to spawn in places that are rather shallow and rich with aquatic vegetation. It has been shown that the spawning beds remain the same each year,[24] and this is a fact that is very important to local fishermen.

The herring feed on plankton, especially saltwater copepods. The study of Rajasilta et al. shows that the salinity conditions of the Baltic Sea highly affect the characteristics of the herring.[25] It seems that due to the decrease in the salinity, the saltwater copepods are disappearing. This development has caused problems with the growth of the Baltic Herring. However, a recent study shows that the herring that spawns in the Archipelago Sea and forage in the Gulf of Bothnia have adapted and learned to feed on the large freshwater copepods (*Limoncalanus macrurus*).[26]

Baltic Herring is an extremely important part of the entity that is here being called the Rymättylä SES. In addition, the geographic area of Rymättylä and its human habitants form another important sub-entity of the local SES.

In the beginning of the twentieth century, Rymättylä was a small municipality with roughly 2,000 inhabitants. The inhabitants relied on the natural resources provided by the Archipelago Sea and the natural environment. The subsistence of the people of Rymättylä relied mainly on fishing and agriculture. The cultivating season ended in autumn and during the harsh winter months the whole Archipelago Sea would usually freeze and be covered with ice until late spring.

For many, winter seining was the only livelihood for half of the year and it was extremely important for the survival of the community. According to ethnologist Veikko Anttila, this form of winter fishing in Rymättylä has a history of at least five centuries.[27] Between 1941 and 1950 winter seining provided the main subsistence for 71 families in the area.[28] During the 1930s, the occupations of the inhabitants became increasingly involved with commercialization, which in the long term influenced the economics of their livelihood profoundly.[29]

Seining is a complex procedure that requires extensive knowledge of the environment, herrings, and the work of nature. For example, the extent and thickness of the ice cover needed to be just sufficient enough to carry the large number of workers, the equipment and the horses or tractors. The Rymättylä fishermen were known not only for their comprehensive knowledge of the movements of the fish, but also of the ice. It was said that the Rymättylä people had made a pact with the devil to be able to walk on such thin ice.[30]

Each seine was owned by approximately eight households, each owning one piece of the net and their share of other equipment. Every winter, after the ice became thick enough, each group of seine fishermen would start their journey in search of the best possible seining spot. Each seine cooperative employed around 50 people – a number that has changed significantly from the 100 employees of the 1880s to the approximately 10 employees of the 1990s – including the owners.[31] There were several different kinds of occupations on the ice but the seine king would be the person in charge of the seine, although it was quite rare that

Figure 9.2 Winter seining equipment. Winter seining fair and exhibition in Rymättylä.

Source: Photo by Kirsi Sonck-Rautio, 2007.

he was one of the owners. Seine kings were the ones who possessed the utmost knowledge of the herring, its movements, and its nature. The seine kings were highly respected members of the community and their knowledge was held to be irreplaceable – until the arrival of the echo sounders in the 1970s.

It is not hard to understand the need for a workforce once one grasps the dimensions of the net. The net itself could be as high as 35 meters and its circumference could be 400 meters. On its own, the net would weigh approximately 600 kilograms.[32] The net was laid under the ice using dozens – or even close to a hundred – holes in the ice, and transported towards the main hole, from which the fish were pulled up with the help of a wooden stick (*uitto*), which was 30 meters long.

The catch was pulled up by several men, using barrels and horses, and later tractors. It was not an easy task, since the catch quite often contained tons of fish. The all-time record for the size of a catch was in 1984, when the local fishermen caught 214 tons of herring. It took six days to harvest them all.[33]

There were several types of celebrations scattered around the seining season, and it dictated the daily, weekly, and monthly schedule of the community for the winter months. Despite the enormous catches of the 1980s, the livelihood soon declined and by the end of the 1990s, there were virtually no winter seiners on the ice anymore. This was not so much due to the lack of herring, but to the lack of ice – the winters in the 1990s were very mild. In addition, the changes in the demographics, the rapid urbanization, increasing competition, and the ageing of the fishermen who had no one to take over their cooperatives[34] had a major influence on the decline of seining as a livelihood.[35]

The Ethnoichthyology of Rymättylä Fishermen

The agency of the herring was apparent in the SES of the old Rymättylä community winter seining. First of all, the herrings' migration routes, homing behavior, and foraging choices all supported and enabled this subsistence mode of winter seining, as well as smaller scale professional or self-subsistence fishing using trap nets or nets. Since it has been proven that the Baltic Herrings' spawning habits and the locations of the spawning beds are very regular, it is fair to argue that a major risk for future generations of fishermen lies in the fact that the herrings' reproduction is highly vulnerable to environmental changes. This is because the eggs of the herrings are often laid in river estuaries and these estuaries are places where environmental disturbances accumulate.[36] The fishermen are aware of this.

Anthropologist Warren T. Morrill, who conducted ethnoichthyological research among the Cha-Cha[37] as early as 1967, defined the local ethnoichthyology as being derived from their fishing methods and the observations their methods required.[38] The Cha-Cha ethnoichthyology is especially concentrated, according to Morrill, on the taxonomy, ecology, and behavior of marine organisms, and the toxicity of fish. Morrill pointed out that although anthropologists have long been interested in folk knowledge, local knowledge regarding marine organisms is one of the most inaccessible, since it is a world to which no human really has access.[39]

The question of taxonomy may not be relevant in the context of northern Scandinavian fishermen, who have been aware of scientific taxonomy for decades, but there are a myriad of different ways of interpreting the characteristics and quality of fish. Many of the fishermen can even recognize a fish that was caught in Rymättylä and define the amount of time that has passed since the herring was harvested, just by looking at it. The method of harvesting is also considered important – fish caught with large commercial trawlers were described as "disgraced."[40]

Similar to the Cha-Cha, Rymättylä people also had a rather extensive knowledge of marine ecology and fish behavior. In Rymättylä, the fishermen have access to scientific knowledge and many of them are in close contact with biologists studying fish in the area. It is apparent though that the views of the local fishermen and those of the biologists are not always similar.[41] In addition, the fishermen occasionally observe phenomena in the area, and they share these observations with biologists.[42] This type of information transfer is referred to as citizen science, a method of collecting data that is becoming more popular among researchers.[43]

The diet of the herring is very important to the local SES, as well as the whole ecosystem of the Baltic Sea. Without proper feeding, herrings do not reach a size that is beneficial to the fishermen. The problem with the herring after the mid-1980s was the reduction in the size of the adult fish. This unwanted development was quickly noticed by the fishermen,[44] and also proven accurate by science. The study of Rajasilta et al. confirmed that the size of the herring was indeed decreasing,[45] thus affecting the fishermen's livelihood as well. However, as an actor in a system, herrings are able to adapt. The herring changed their diet to fresh water copepods, thus influencing the functions of the whole SES.

The importance of herrings as regards the local SES cannot be limited to economic and ecological sectors alone. It had a vast influence on the socio-cultural sphere, and therefore the SES analysis including herring agency is a very important tool in promoting socio-cultural sustainability. The fishermen in Rymättylä noted that in a way, herrings democratized the whole community. The cooperative nature of the livelihood created the sense of all being equal, but also because the ecological knowledge and skills related to it were accessible to everyone:

> If you were skilled, like those seine kings or others, to be a respected part of the community you wouldn't have to be a land owner … we all had dignity and value. It wasn't like the land owner and tenant farmer, but this type of fishing sort of made us equal within the community.[46]

Besides the sense of equality, the winter livelihood provided by the herring also influenced the way people in the community interacted and socialized together. Although winter seining brought people together, it also separated them, since different cooperatives were in competition with each other. The knowledge of excellent harvesting spots was kept secret and groups were half seriously suspicious of each other:

> [And] in winter we would just spy on each other and be competitive, but in spring, when the seine was led to winter rest, those quarrels were bygones. … Those [quarrels] stayed on ice. Sometimes someone would refer to some disagreement and we'd be like "it was back in winter."[47]

Winter seining influenced the life of the whole community, not only the fishermen themselves. During the summer, the social life diminished and fishermen became farmers. It was quite common even in the 1960s, that boys – and girls – would leave school and start working in the cooperatives when they turned 15 years old. Being able to go on the ice and participate in men's work was something that boys looked forward to.[48] Even the church had to follow the schedule dictated by winter seining. Most Finnish children take their Evangelic-Lutheran confirmation at the age of 15, which required studying and an examination. These exams, however, would be postponed until spring, due to the winter seining.

After the 1960s, the trawling business started to appear. This was seen as a negative thing, both micro-economically and ecologically. In addition to the mild winters and declining amounts of ice, the trawlers were seen as the main reason for the "extinction" of the winter seining. This more effective way of catching herring was accused of distorting the market. Some fishermen also thought that trawlers disturbed the stocks in the open sea, thus preventing the fish from ever reaching the coastal areas of the archipelago. The herring caught with seine were seen – among the old winter seining fishermen – as proper herring. Winter seining was perceived as a more sustainable way of fishing, since there were even written rules, "the Seine law," which dictated, among other things, the number of times each cooperative could go on the ice. It seems, however, that this law was not written with the sustainability of the fish stock in mind, but to ensure the fairness of the game, so to speak.[49]

Despite the decline in the economic importance of herrings, their importance still exists in a traditional and cultural sense. Fishing – mainly as a leisure or self-subsistence activity – is still visible and considered to be a way of life.[50] Many local people, regardless of not being fishermen themselves, wish to maintain this old tradition of fishing as a livelihood. For example, each year the local Lions club organizes an annual winter seining fair where – if the ice conditions allow – they arrange a winter seining demonstration, among other things.

The importance of the herring can be demonstrated by the fact that in front of the former town hall stands a statue of winter seiners and even the coat of arms of Rymättylä includes a representation of two herrings.[51]

Interestingly enough, after taking the photo of the coat of arms, I was told that it is rarely acknowledged that the fish in it are not herring at all. Apparently, the picture of a random species of an unspecified grey fish was used and everyone just assumed it was a herring.[52] This collective local misinterpretation indicated how deep the connection between herring and Rymättylä is – it even creates optical illusions.

Conclusion

It has been calculated that all the wild fish stocks in the world will have collapsed by 2048.[53] The severity of the collapse has not been agreed by everybody, but even so, the demise of fisheries is something on which most researchers have reached a consensus.[54] In addition, the importance of the socio-cultural perspectives of fisheries management has finally been widely recognized, and studies show that fishermen's folk knowledge of fish reproduction, migration, and diet very often compliment scientific data.[55] In researching adaptation strategies, conservation policies, and sustainable fisheries management, there are no better experts than the herrings themselves, and the fishermen who possess LEK regarding the fish and the environment. Their knowledge should not be treated as knowledge that can merely complement scientific biological data or research, but should also be considered the other way around.

The case of Rymättylä is one example of a fishing community where the fishermen had a tremendous amount of LEK, which included the understanding of the agency of the herring as a whole, the meaning of this specific species to the whole community, and how the changes in the herring's own life change the system. The fishermen were aware both of their environmental surroundings and the special circumstances related to their livelihood, as well as ways of managing fisheries in traditional ways that were both culturally, socially, and, to some degree, ecologically sustainable. Economically, the system had been able to remain somewhat sustainable for centuries, but the traditional SES started to change drastically when it became engaged with commercialization in the 1930s, and by the 1960s the self-sufficient nature of the livelihood was merely a memory. With the pressure of competition and globalizing markets the foundations of the system started to transform.

Nevertheless, since the role and agency of the fish is so crucial to the fishing industry and the local SES connected to it, it should be highly beneficial to include the fish in the analysis, not merely as a natural resource, but as an active participant, as an actor. The agency of the herring was apparent in the SES of the old winter-seining Rymättylä community. Their migration routes, homing behavior, and foraging choices all supported and enabled the subsistence mode of winter seining, as well as the smaller scale professional or self-subsistence fishing by trap net or nets. The relationship between local people and the fish were multi-layered and interconnected. Cultural and social sustainability is hard to reach if ecological and economic sustainability are absent. However, I argue that ecological and economic sustainability at a local community level are more possible to attain if socio-cultural sustainabilities are also included in management plans. It is crucial to nurture LEK in order to maintain knowledge about animal agency. When planning, it is important to remember that anything planned does not just happen to the fish; the fish just might have a mind of their own.

Notes

1 Rymättylä merged with the city of Naantali in 2009.
2 Sonck, "'Se on Rymättylän.'"
3 I conducted fieldwork in Rymättylä in 2007 and 2015–2016. Among the informants I have 16 fishermen, their level of professionalism varying from self-subsistence to commercial professional fisherman, aged between 45–85. When quoting I will refer to informants by their code number (M = male and F = female).
4 McFarland and Hediger, *Approaching the Agency*, 1.
5 Sumpter et al., "Consensus Decision Making by Fish," 1773–7.
6 Reebs, "Can a Minority of Informed Leaders," 403–9.
7 Ibid., 403; Mazeroll and Montgomery, "Structure and Organization of Local Migrations," 89–106.
8 Couzin, "Collective Cognition in Animal Groups," 36.
9 Ibid., 41; Reebs, "Can a Minority of Informed Leaders," 407–8.
10 Killen, Nati, and Suski, "Vulnerability of Individual Fish."
11 Latour, *Reassembling the Social*, 46.
12 Ibid.,53.
13 Hegland, "Fisheries Policy-Making," 219–37.
14 Hansson et al., "Managing Baltic Sea Fisheries," 265–71.
15 McClanahan et al., "Healing Small-Scale Fisheries," 33-47; Huntington, "Using Traditional Ecological Knowledge," 1270–4; Zykowski, Curtis, and Watts, "Using Fisher Local Knowledge," 120–7; Gadgil et al., "Exploring the Role," 189–208.
16 Gallopín, "Linkages between Vulnerability," 293–303.
17 Smit and Pilifosova, "Adaptation to Climate Change," 879.
18 West and Brockington, "An Anthropological Perspective on Some Unexpected Consequences of Protected Areas," 609–16.
19 Cruikshank, *Do Glaciers Listen*?, 9.
20 Huntington, "Using Traditional Ecological Knowledge," 1270.
21 Ibid., 1273; Zykowski, Curtis, and Watts, "Using Fisher Local Knowledge," 120–7; Gadgil et al., "Exploring the Role," 189–208.

22 Natural Resource Institute of Finland, "Silakka."
23 Kääriä et al., "Spawning Bed Selection," 917–23; Rajasilta, "Timing of Spawning"; Rajasilta et al., *Intensive Monitoring of Spawning Populations*.
24 Kääriä, "Reproduction of the Baltic Herring"; Rajasilta et al. *Intensive Monitoring of Spawning Populations*, 53.
25 Rajasilta et al., *Intensive Monitoring of Spawning Populations*, 60–1.
26 Hänninen, "Aikasarjatutkimukset Itämeren tulevaisuuden avaimena," 78.
27 Anttila, *Rymättylän nuottakalastus muuttuvana elinkeinona*, 3.
28 Matinolli, *Rymättylän historia*, 316; Sonck, "'Se on Rymättylän,'" 24.
29 Sonck, "'Se on Rymättylän,'" 37.
30 Field journal 2007.
31 Sonck, "'Se on Rymättylän,'" 43.
32 Matinolli, *Rymättylän historia*, 253.
33 Sonck, "'Se on Rymättylän,'" 40.
34 Ever since the rapid motorization, the need for workforce had diminished, and fishermen with proper know-how were rare.
35 Sonck, "'Se on Rymättylän,'" 78–84.
36 Rajasilta et al., *Intensive Monitoring of Spawning Populations*, 3.
37 The term used to refer the people of St. Thomas Island with French origins.
38 Morrill, "Ethnoichtyology of the Cha-Cha," 405, 411.
39 Morrill, "Ethnoichtyology of the Cha-Cha," 407.
40 Field journal 2015.
41 Field journal 2015.
42 Interview 2015, Informant M9.
43 Silvertown, "A New Dawn for Citizen Science," 467–71.
44 Field journals and interviews 2007 and 2015–16.
45 Rajasilta et al., *Intensive Monitoring of Spawning Populations,"* 65.
46 Interview 2007, Informant 6M. See also Sonck, "'Se on Rymättylän,'" 47.
47 Interview 2007, Informant 3M.
48 Sonck, "'Se on Rymättylän,'" 50.
49 Sonck, "'Se on Rymättylän,'" 46.
50 Field journal 2007.
51 Field journal 2007.
52 Field journal 2015.
53 Worm et al., "Impacts of Biodiversity Loss," 787–90.
54 Hilborn, "Moving to Sustainability," 296–303.
55 Silvano and Begossi, "Ethnoichthyology and Fish Conservation," 285–306; Silvano et al., "When Does This Fish Spawn?," 371–86; Rebelo, Freitas, and Soares, "Fish Diet from Manacapuru," 39–44.

Bibliography

Ethnographic Materials

All field journals, audio, and transcripts of the interviews are in the possession of the author.

Literature

Anttila, Veikko. *Rymättylän talvinuottakalastus muuttuvana elinkeinona 1885–1967*. Turku: University of Turku, 1968.

Couzin, Iain D. "Collective Cognition in Animal Groups." *Trends in Cognitive Science* 13 (2009): 36–43. doi:10.1016/j.tics.2008.10.002.

Cruikshank, Julie. *Do Glaciers Listen? Local Knowledge, Colonial Encounters and Social Imagination*. Vancouver: UBC Press, 2005.

Gadgil, Madhav, Per Olsson, Fikret Berkes, and Carol Folke. "Exploring the Role of Local Ecological Knowledge in Ecosystem Management: Three Case Studies." *In Navigating Socio-Ecological Systems: Building Resilience for Complexity and Change*, edited by Fikret Berkes, Johan Colding, and Carl Folke, 189–208. Cambridge, UK: Cambridge University Press, 2008.

Gallopín, Gilberto C. "Linkages between Vulnerability, Resilience, and Adaptive Capacity." *Global Environmental Change* 16 (2006): 293–303.

Hänninen, Jari. "Aikasarjatutkimukset Itämeren tulevaisuuden avaimena." *Seili. Saaristomeren tutkimusta 50 vuotta*, 69–80. Archipelago Research Institute Publications. Turku, Finland: Suomen Yliopistopaino, 2014.

Hansson, Sture, Olle Hjerne, Chris Harvey, James F. Kitchell, Sean P. Cox, and Timothy E. Essington. "Managing Baltic Sea Fisheries under Contrasting Production and Predation Regimes: Ecosystem Model Analyses." *AMBIO: A Journal of the Human Environment* 36 (2007): 265–71.

Hegland, Troels J. "Fisheries Policy-Making: Production and Use of Knowledge. The Knowledge Base for Fisheries Management." *In Developments of Aquaculture and Fisheries Science*, vol. 36, edited by Lorenzo Motos and Douglas Clyde Wilson, 219–237. Amsterdam: Pergamon Press, 2006.

Hilborn, Ray. "Moving to Sustainability by Learning from Successful Fisheries." *A Journal of the Human Environment* 36 (2007): 296–303.

Huntington, Henry P. "Using Traditional Ecological Knowledge in Science: Methods and Applications." *Ecological Applications* 10 (2000): 1270–4.

Kääriä, Juha. "Reproduction of the Baltic Herring (Clupea harengus membras L.): Factors Affecting the Selection of Spawning Beds in the Archipelago Sea, in SW Finland." PhD diss., University of Turku, 1999.

Kääriä, Juha, Marjut Rajasilta, Mika Kurkilahti, and Martti Soikkeli. "Spawning Bed Selection by the Baltic Herring (Clupea harengus membras) in the Archipelago of SW Finland." *ICES Journal of Marine Science* 54 (1997): 917–23.

Killen, Shaun S., Julie J.H. Nati, and Cory D. Suski. "Vulnerability of Individual Fish to Capture by Trawling Is Influenced by Capacity for Anaerobic Metabolism." *Proceedings of the Royal Society B* 282 (2015). doi: 10.1098/rspb.2015.0603.

Latour, Bruno. *Reassembling the Social. An Introduction to Actor-Network-Theory*. Oxford: Oxford University Press, 2005.

McClanahan, Timothy R, Juan Carlos Castilla, Alan T. White, and Omar Defeo. "Healing Small-Scale Fisheries by Facilitating Complex Socio-Ecological Systems." *Reviews of Fish Biology and Fisheries* 19:1 (2009): 33–47. doi: 10.1007/s11160-008-9088-8.

McFarland, Sarah E. and Ryan Hediger. "Approaching the Agency of Other Animals: An Introduction." *In Animals and Agency: An Interdisciplinary Exploration*, edited by Sarah McFarland and Ryan Hediger, 1–20. Boston: Brill, 2009.

Matinolli, Eeva. *Rymättylän historia. II.I: Luonto, 1800- ja 1900-luvut, murre*. Turku, Finland: Rymättylän kunta ja seurakunta, 2000.

Mazeroll, Anthony I., and W. Linn Montgomery. "Stucture and Organization of Local Migrations in Brown Surgeonfish (Acanthurus nigrofuscus)." *Ethology* 99 (1995): 89–106.

Morill, T. Warren. "Ethnoicthyology of the Cha-Cha." *Ethnology* 6 (1967): 405–16. doi: 10.2307/3772827.

Natural Resource Institute of Finland. "Silakka (*Clupea harengus membras*)." Accessed June 9, 2014. http://rktl.fi/kala/tietoa_kalalajeista/silakka.

Rajasilta, Marjut. "Timing of Spawning in the Baltic Herring (Clupea harengus membras) in the Archipelago Sea, SW Finland: Regulatory Mechanisms and Consequences for Offspring Production." PhD diss., University of Turku, 1992.

Rajasilta, Marjut, Jan Eklund, Päivi Laine, N. Jönsson, and T. Lorenz. *Intensive Monitoring of Spawning Populations of the Baltic Herring (Clupea harengus membras L.)*. Turku, Finland: SEILI Archipelago Research Institute, 2006.

Rebelo, Sergio Roberto Moraes, Carlos Edwar de Carvalho Freitas, and Maria Gercilia Mota Soares. "Fish Diet from Manacapuru Big Lake Complex (Amazon): An Approach Starting from the Traditional Knowledge." *Biota Neotropica* 10 (2010): 39–44.

Reebs, Stephan G. "Can a Minority of Informed Leaders Determine the Foraging Movements of a Fish Shoal?" *Animal Behaviour* 59 (2000): 403–9. doi: 10.1006/anbe.1999.1314.

Silvano, Renato A.M., and Alpina Begossi. "Ethnoichthyology and Fish Conservation in the Piracicaba River (Brazil)." *Journal of Ethnobiology* 22 (2002): 285–306.

Silvano, Renato A.M., P.F.L. MacCord, R.V. Lima, and Alpina Begossi. "When Does This Fish Spawn? Fishermen's Local Knowledge of Migration and Reproduction of Brazilian Coastal Fishes." *Environmental Biology of Fishes* 76 (2006): 371–86.

Silvertown, John. "A New Dawn for Citizen Science." *Trends in Ecology & Evolution* 24 (2009): 467–71. doi:10.1016/j.tree.2009.03.017.

Smit, Barry, and Olga Pilifosova. "Adaptation to Climate Change in the Context of Sustainable Development and Equity." *In Climate Change 2001: Impacts, Adaptation, and Vulnerability*, edited by James J. McCarthy, Oswaldo F. Canziani, Neil A. Leary, David D. Dokken, and Kasey S. White. Cambridge: Cambridge University Press, 2001.

Sonck, Kirsi. "Se on Rymättylän niin vanha perinnehomma toi ammatti" – *Ympäristön, yhteiskunnan ja paikallisyhteisön vuorovaikutus Rymättylän talvinuottauksen muutoksessa 1880-luvulta 1980-luvulle."* Master's thesis, University of Oulu, 2011.

Sumpter, David J.T., Jens Krause, Richard James, Iain D. Couzin, and Ashley J.W. Ward. "Consensus Decision Making by Fish." *Current Biology* 18 (2008): 1773–7. doi:10.1016/j.cub.2008.09.064.

West, Paige, and Dan Brockington. "An Anthropological Perspective on Some Unexpected Consequences of Protected Areas," *Issues in International Conservation* 20 (2006): 609–16.

Worm, Boris, Edward B. Barbier, Nicola Beaumont, J. Emmett Duffy, Carl Folke, Benjamin S. Halpern, Jeremy B. C. Jackson, Heike K. Lotze, Fiorenza Micheli, Stephen R. Palumbi, Enric Sala, Kimberley A. Selkoe, John A. Stachowicz, and Reg Watson. "Impacts of Biodiversity Loss on Ocean Ecosystem Services." *Science* 314 (2006): 787–90. doi: 10.1126/science.1132294.

Zykowski, Sylvia, Allan Curtis, and Robyn J. Watts. "Using Fisher Local Ecological Knowledge to Improve Management: The Murray Crayfish in Australia." *Fisheries Research* 110 (2011): 120–7. doi:10.1016/j.fishres.2011.03.020.

Part III

Agency Represented

10 *Rangifer Tarandus* in Textual Service

Positioning Reindeer as an Agent in a Human/Non-Human Interactional Context in Three Texts

Jukka Nyyssönen

Introduction

This chapter is about ways in which agentic reindeer (*Rangifer tarandus*) serve identity politics in three texts. The texts studied are *Iter Lapponicum* by Carl von Linné, *Antropogeografiska studier inom Petsamo-området, 1, Skoltlapparna* (Human Geographical Studies in the Petsamo-Region 1, Skolt Lapps) by Väinö Tanner, and *Suenjelin saamelaisten perintö* (The Legacy of Suenjel Sami) by Matti Sverloff.[1] All three authors/scholars have included passages of reindeer in an interactional context with humans (Sami in all cases) in their books.

This chapter is also part of a new theorization on the disappearance of the division between culture and nature, not necessarily by technologizing nature or through its disappearance,[2] but by natural, living beings entering and defining human cultures. This is achieved by acknowledging animal agency and its limits, that humans and animals are not only interacting but also shaping one another's behavior, and that animals are capable of action, which has an effect on human agency.[3] Recent tendencies in other branches of science have taken as their starting point the premise of cognitive ethology, that animals manifest many cognitive capacities and each (individual) capacity of an animal should be approached from that animal's viewpoint. One finds oneself in the radical corner of the scholarly field, part of a critique of anthropocentric attitudes and ideas of human exceptionalism that are built on the belief in a strict division between humans and non-humans. This critique builds on the notion of nature as an agentic force in interaction with other beings, including humans.[4]

Attempts have been made to define animal agency separately from human parameters, with varying results. Definitions of animal agency are readily centered on terms that describe human agency (e.g., morality, emotions, free will), but Sarah McFarland and Ryan Hediger stress that animals are agents in their own ways: they act and react in ways that are standard to their own group and they have their own goals and motives. The "meaning" and "intention" of animal communication and reactions are on the far side of communicative[5] and cross-cultural[6] barriers, ultimately a matter of speculation and a source for misinterpreting.[7] As there are obvious and severe epistemological difficulties in deciding upon levels of individuality, consciousness, moral sense, free will and autonomy among non-humans, agency is found at an interactional, animal-to-animal level: in the capacity of non-humans to sustain communicative systems, as well as in their capacity to act, operate and exert power in order to achieve certain goals. The capacity of directed action to sustain oneself as well as acting on threats and perceptions of prey is central in attributing animal agency.[8]

Non-humans have no understanding of abstract constructs and structures surrounding and limiting their lives, such as the capitalist food market, game administration, etc. The non-humans meet these limits when they are enforced by humans,[9] and then they can engage in context-specific action, for example resistance. Household and herded animals, and those kept in captivity, can defy human supremacy and conditions, experienced in their closest context by numerous means, including acts of killing.[10] On the other hand, in the own subjective universe of an animal, or Umwelt as Jakob von Uexküll terms it, objects are perceived and responded to according to the functional signs or tones they emit for each individual subject. It is in the relations embedded in other animal Umwelts, the overlaps between them, and the human vicinity, the human/non-human interactional context, where some aspects of reindeer agency become observable, making interspecies communication (to some extent) possible. Human vicinity provides reindeer with protection, relief, nutrition and *affordances* (introduced later in the text) of various sorts that can be utilized by the reindeer, and which are based on the reindeer's choice, sensory and cognitive capacities.[11]

Animal agency is to be found, on the other hand, in the positions and roles that animals fulfill and the action they take in networks binding humans and non-humans. Agency is also found in the historical, economic, social, psychological and identificational significance attributed to animals.[12] The main focus of this chapter is the identificational aspect, the meanings attached to the agentic reindeer in three texts, connected by intertextual ties. I study ways in which the agentic reindeer is utilized in scholarly production as a *rhetorical animal*. I study representations of her agency and ways she deals with humans as part of the identity politics in the three chosen texts.[13] As a representation, as the depicted and the signified, the animal conveys meanings to humans through the meaning and subject position attributed to her.[14] What meanings are attached to reindeer agency, and what kind of agenda is it intended to serve in the texts? Which folk groups are defined using reindeer, and how?

Carl von Linné's Sense of Awe before the Agentic Animal

The context of writing for the famed naturalist and author of *Systema Naturae*, Carl von Linné (1707–1778) was a utilitarianist-mercantilist-colonial-economic thrust, seeking natural resources in order to accumulate national wealth. In the seventeenth and eighteenth centuries, expeditions were coupled with scientific interest and exoticism. Swedish expeditions were utilitarian in their search for new plants for commercial usage, coupled with a mercantilist economic policy and a belief in hidden treasure in the terrain of Sweden bestowed as a blessing from God. von Linné presented his 1732 expedition to the Swedish north first and foremost in terms of its potential for accumulating scientific knowledge, but the secondary grounds were the accumulation of national wealth. The expeditions and reports on different aspects of flora, fauna, history and minerals, etc. encountered on these journeys were a part of the academic requirements of the time.[15]

Having said this, von Linné's sensitivity concerning the use of the Sami as informants has not gone unnoticed by researchers. von Linné's travels took place during a shift in the gaze upon the Sami from a fearful and demonizing mythic/religious gaze upon a wild and bestial people capable of witchery to the curious gaze of a natural scientist trying to find out how these people survived in their harsh surroundings; the latter resulted in mostly positive representations of Sami pastoral capabilities.[16] Nor was von Linné's gaze unreceptive to the agentic aspects of the reindeer, as we shall see.

Aside from *Iter Lapponicum*, von Linné produced other texts on reindeer based on his observations during his trip to Lapland in 1732. One of them is "Om Renarnas Brömskulor i Lapland" ("On Reindeer's Gad-Fly Sores in Lapland," 1739). The text is concerned with the plague of the caribou warble fly (*Hypoderma tarandi*, or *Oedemagena tarandi*, *lávžá* in northern Sami) on the reindeer.[17] There is a passage depicting ways of dealing with this plague. The Forest Sami left the forests for the mountains because of the flies. von Linné had visited some Sami who were still residing in the forests, even though the plague was at its worst, and the growing antlers of the reindeer were bleeding because of biting flies (though not the warble fly). von Linné describes how, in order to cope with this, the reindeer sought out a hut where the Sami had set light to some bracket fungus (von Linné calls it *Agaricus abietis*, possibly *Fomitopsis pinicola*) and the thick smoke generated from the fire gave the reindeer a moment of peace and relief from the ongoing plague. The reindeer, he reports, "knowing the effect (of the smoke), lie down and sleep."[18] He also depicts the way in which the herd reacted to the presence/humming of the flies by lifting their heads and "kicking with their feet" in unison and being constantly on the move.[19]

von Linné further depicts the way in which the reindeer travel against the wind to rid themselves of this plague. In the eyes of von Linné, this created a conflict of interest: the Sami wanted the reindeer to travel in the direction of the wind, making it easier to herd them. This conflict was deepened by the way in which von Linné felt that the Sami were unaware of the relationship between the fly and the bites (Sami: *curbma*) in the reindeer skin, indicating the presence of the parasite larva.[20] The way in which reindeer had "learned" to fear the fly and to disobey the herder, in order to avoid the plague, astounded von Linné; so too did the way in which the flies followed the reindeer for days, exhausting themselves many times, in order to plant their eggs. In a physico-theological spirit, von Linné credits much of this to the work of a fourth actor in the interrelational context, the Creator,[21] but also in two other directions: he credits the little insect with the major migrations from the woods to the mountains undertaken by the Sami in order to cope with the flies; in his diaries, he also acknowledges a non-human capacity for learning.[22]

In addition to reporting simultaneously on two different sets of insects, the depiction of reindeer herding and Sami knowledge is sometimes self-contradictory and unconvincing: in his 1732 diary, von Linné depicts the reindeer "coming home" to be milked, but realizes the day after that the herd is actually herded to the *lavvo* by a young girl and her dog (von Linné depicts the high level of control by which the herd is mastered by the two of them).[23] The depiction of the Lapp village in the mountain presented in *Iter Lapponicum* includes elements of herd control, as well as depictions of the proximity of the Sami to the stock. von Linné wondered how the Sami could tell which reindeer had been milked from the vast herd. An informant told him that the Sami had a name for each reindeer and knew them all individually. von Linné was impressed by this, admitting that he could not tell one reindeer from another. Although there are highly normative passages in the text concerning the Sami (their lack of agricultural and capitalistic rationality, their tendency towards lower morals and hygiene, etc.), von Linné also demonstrates a capacity for wonder and enquiry, not simply condemning Sami rationality as insufficient. In the end, though, his gaze is more stringent upon the Sami than the reindeer: the actions of the reindeer are listed to their credit, but those of the Sami pulled them on both directions, resulting in the categorization of the Sami in the intermediate category of *monstrosus*.[24]

von Linné also depicts other aspects of reindeer agency: a tendency to run away from the herder and to gather in herds, in order to deal with the mosquitoes: "they like to be together – the more of them there are, the better," a factor also used by the herder when gathering the herd. He also points out the potential human contribution in the reindeer social sphere, e.g., protection from wolves.[25] What is of particular interest here is the way in which von Linné credits agency (e.g., maintaining a distance from the herder) and cognitive capabilities (e.g., using the landscape and human assistance in order to cope with irritations) to the non-human in a human/animal context, factors which he frequently ascribed to other beings as well.

The way in which humans and animals are treated and categorized in the same manner, as part of their environment, and seemingly under the same premises, was a convention of categorizing natural sciences at that time. Humans, even the *monstrosus*, were not labeled as animals, even though the effect might appear so, and even though von Linné credited animals with cognitive capabilities and even a soul, an idea for which he was severely criticized.[26] There are also traces of an older tradition, based on the aforementioned capacity to experience wonder in observing non-humans and ponder the purpose of these beings in the world. In earlier times, and in cultures outside those of the Judeo-Christian/Aristotelian, hierarchies were more fluid, not strictly placing humans at the top,[27] but von Linné perfected the other tradition, that of scientific taxonomy and classification, which naturalized the place of humans at the very top in a competing taxonomic system.[28]

Being part of nature or the human in a human/non-human context could push humans in both directions, down and up the ladders of the increasingly racial hierarchies of that time. Nomads have never ranked high in these hierarchies: contact with animals is not in their favor. In von Linné's case, the resulting *monstrosus* categorization pointed downwards on the scales reflecting levels of culture and civilization.[29] In Väinö Tanner's scheme, the direction and aim of representation appear to achieve the opposite.

Tanner and Reindeer as the Key to Alternative Rationalities

Geologist and professor of geography at the University of Helsinki, Väinö Tanner (1881–1948) became acquainted with reindeer and herding as the chairman of pasture committees working on the pasture crisis in Torne Lappmark in the northeasternmost Swedish Sami villages in the early twentieth century. During fieldwork, he spent some weeks trekking the terrain. Besides gnawed pastures, committee members encountered only a few reindeer, which were residing in other pasture regions during their stay. There is one depiction of an interspecies encounter, as a lone reindeer ox joined the commission at their camp on August 26, 1910. The Sami guide and informant Per Persson Maraven was able to inform them that the ox belonged to Forest Sami Isak Isaksson Viihusjärvi. Tanner was quite fascinated by the ox, which followed the committee for three days. He took the opportunity to observe its grazing habits with his "Lapp-advisors" Valle and Aikio.[30] According to information from this source, the reindeer ox is so passive, and the information is derived from such mundane aspects of being a reindeer, that it may not qualify as a co-constructor of knowledge (see Nora Schuurman in this volume), but rather as a source of information and an object of observation. Positioning of the reindeer in the ownership and property networks highlights the passive role of the reindeer as well.[31] Tanner took the insights he learned from reindeer, their herding and environment, and the ideal form of herding, which he internalized, and worked on them further in his seminal work on the Skolt Sami.

Tanner visited Petsamo, in northernmost Finland, in his capacity as a state geologist in search of nickel ore, but Tanner took the expeditions to another direction. He spent the summers in Petsamo with Skolt Sami assistants. The fieldwork and texts by von Linné on reindeer and their parasites, stored in his archive, deepened his view of the human–reindeer relationship. Tanner discovered the agentic reindeer and how this affected human behavior. According to draft versions of his texts, the biology of the fish decided the whereabouts of the half-nomadic people during the year, while reindeer biology and choice of pasture dictated the same for the full nomads. Tanner had omitted the potentially pejorative connotation of a people following the animal, and in doing so he simultaneously raised fish and reindeer as agents with interests; in his eyes they defied the romantic notions of freedom often associated with the Sami.[32] This constituted one of the starting points in Tanner's synthesis of Skolt Sami subsistence, where the adaptation, which he coded the most rational, occurred to the geographical surroundings in a way that secured aspects of agency to all parties: non-humans, the landscape and the semi-nomadic herder/fisherman.

Tanner wrote about the reindeer–human relationship when reindeer are taken to a new pasture ground: the reindeer tend to escape back to the previous pastures, especially during the spring, and disappear from the owner.[33] Rather than only to resist human aims and avoid human contact, Tanner, like von Linné, went further in his analysis of reindeer acting in the vicinity of humans in a way that resembles what was later to be known as human–animal context networks.

These networks enable researchers to chart a space for action and the possibility of semi-domesticated animals entering willingly into, or withdrawing from, human association and the *affordances* offered thereof. Affordances recognize animals' abilities of discernment and decision-making, such that they respond to and act upon reciprocating animals and objects in their surroundings that are deemed significant. Affordances are opportunities for action, a fact of environment and of behavior; they are resources one perceives in an environment. Affordances are interactional phenomena that may in turn evoke diverse responses, depending upon the situation and the individuals involved. The perception of affordances is thus shaped not only by the sensory capabilities of the animal, but also by an individual's awareness, memory and understanding of previous experiences.[34]

Tanner reported the reindeer's willingness to dwell in the social sphere, in the vicinity of humans; he described their habit of seeking shelter, of their own accord, from the heat and the mosquitoes in a *liahka* (reindeer barn) in the Skolt Sami *siida*'s summer herding.[35] Tanner wrote:

> Most of the (Skolt Sami) families allowed their reindeer to go free during the summer. Others, in their turn, herded their reindeer carefully in the vicinity of their summer residence/fishing ground, so that the domestication of this usually half-wild animal had reached so to say its peak. For the well-being of the reindeer they had erected a 20-metre building resembling a long, turf-roofed shelter called *liáhka*, in which the reindeer sought shelter from the blazing sun and the mosquitoes. Of their own accord, the animals went to the shelter around seven o'clock in the morning, as the sun began to warm the air. Having rested, the reindeer went out at around seven o'clock in the evening to "take a stroll" and graze during the night.[36]

For reindeer, the mosquito plague provides a means of relating to the affordances offered humans in their landscape. Most usually reindeer were attracted to the fires lit by the herders, and gathered to be milked and to obtain relief from the mosquitoes.[37]

In the interpretative and discursive field, which Tanner constructed, there was also an instrumental aspect: the reindeer and the human/non-human dynamics were a factor in different ethnic groups being represented and placed in hierarchies. Unlike von Linné, Tanner used the reindeer to raise the formerly "savage" herder up the ladders of culture and rationality. Tanner constructed a long cultural continuity and presence, but most importantly a viable, *best adapted subsistence form*, which he coded in the Sami's favor, signaling their choice of optimum adaptation to the environment.[38] While many contemporaries of Tanner defined the Sami as an irrational folk group wandering around aimlessly in the mountains, Tanner constructed an opposing scheme based precisely on reindeer habits, an annual cycle of seeking pastures, in an interaction with the herder making planned interventions in the reindeer Umwelt.

The Social Darwinistic scales were more refined than those of von Linné: reindeer herding was a substitute for hunting[39] and could be used in imperialistic projects in order to raise the herders up the ladders of civilization. In Tanner's archive there are notes on the literature dealing with projects to introduce reindeer herding to the Inuit people in North America, in Alaska.[40] Tanner had noted the idea by J. L. Underwood of "reindeer as a civilizer." On many occasions the motive beneath layers of patronizing and racist attitudes was to raise people dependent on hunting subsistence up to a more secure livelihood. What had not changed and what Tanner could not escape either was that connectedness with nature still was a mark of primitiveness that placed the Sami lower in the cultural evolution, often accompanied by aggressive representations of laziness, immorality, etc.[41] To Tanner's credit, he coded the Skolt Sami herding differently. His depiction of herding stressed the warmth and closeness of the relationship as much as the skills, capabilities and the expertise of the Skolt Sami – here Tanner leans towards an older way of representing the Sami, such as that of von Linné.

The identity politics of the book are complicated, not solely interested in the rehabilitation of the Sami; the book is unfinished and the stress laid on adaptation to the reindeer economy complicates the agenda of Tanner. The associated significance of the reindeer was politicized by Tanner in the discursive fields of academia, but the built-in naturalized hierarchies of the time meant that Tanner's mission to lift the Sami was partially doomed to fail: this is evident in Tanner's tendentious discussion about ways in which Scandinavian and Finnish settlers adapted to reindeer herding. The Scandinavians adopted it successfully, as part of their farming economy, while the Finns assimilated the Sami ways (with the connotation of collapsing downwards).[42] In Tanner's scheme there had to be something lower than the Scandinavians and Finns and that was the Sami. The scheme demonstrates the strength of the discourses of that time, which disrupted Tanner's identity political agenda concerning the Sami. It was not only the rhetorical animal, and the ways in which people had chosen to adapt to this animal, that pinpointed their place in the hierarchy.

Matti Sverloff and the Compassionate Reindeer Revitalizing a People

Matti Sverloff enjoyed a long career as an ombudsman for the Skolt Sami, and as a Sami politician and activist. His book on the Suenjel Sami is more of a pamphlet than a scholarly work. With direct reference to Tanner, Sverloff continued his quest for the rehabilitation of the Skolt Sami and deepened the discussion of the human/non-human interactional context. A trope detectable in both Tanner and Sverloff's work – the closeness and warmth experienced in the relationship between human and animal – becomes more central on Sverloff's text, revealing an effort to utilize the environmentalist discourse of our time by representing the exceptional closeness between the Skolt Sami and reindeer.

Matti Sverloff represents the old Skolt Sami herding as highly intensive, with reindeer roaming in very close contact, within the fence near the home of the herder, very tame for most of the herding year. The fences were used until 1928, but even after that the reindeer remained in the vicinity of people. Sverloff stresses the role of the individual reindeer in his depiction of Suenjel reindeer herding, resisting the taming process. The reindeer demonstrated many kinds of resistance, according to its character, including stalling and kicking.[43]

In addition to the individual reindeer, Sverloff wished to show a deeper understanding of the dynamics within the stock and the independence of each animal. Sverloff credited the reindeer with personal characteristics (given by the herder, who knew each reindeer personally) and also the capacity to act against predators in unison. He also credited the reindeer, or the older reindeer ox, for sentiment and solidarity with the reindeer fawns: during winter digging, the older ox dug holes for the smaller ones to eat; they "fed the smaller ones," showing thus "good manners," as Sverloff writes.[44]

The "good manners" mentioned by Sverloff may be an attributed agency, or an anthropomorphism.[45] The sources are bound to contain both categories: *anthropomorphisms* attributing meaning to animal behavior according to human parameters, and more direct reports of observations, approximating what Kari Weil calls *critical empathy* grounded in recognition of difference and inaccessibility. On the other hand, anthropomorphism may be seen as a strategy of writing/representing inventively about the "Other," escaping the limitations of the Umwelt and the situatedness of one's own environment.[46] Attributing reindeer with compassionate sentiment was not unique to Sverloff. One informant of Yngve Ryd told how reindeers can become acquainted with one another and become friends on a one-to-one basis and start to graze together. Johan Turi credited the cows with sentiment: they are easy prey for the wolves after the latter have killed the calves, since the cows simply run around in grief, "because they have lost their children."[47] This concept may be both a depiction based on observation *in situ* and an anthropomorphism, a literary technique chosen to make the non-human animal more accessible to humans.

Matti Sverloff's representation of seamless co-operation is not purely centered on reindeer capacities. Sverloff is more interested in representing Skolt Sami relations with the reindeer as exceptionally deep, as well as focusing on the craft and high level of expertise in handling the reindeer. Sverloff's book has an instrumentalist side to it: it was meant to serve as a rehabilitation of the Skolt Sami, the Sami group considered in many ways to be the lowest of the Sami groups in Finland. This image was cultivated in self-representations as victims, based on a long history of being stigmatized as reindeer thieves, people of low morals and people foreign to Finnish ways. The reindeer as an actor in the discursive universe of the book serves one of the instrumentalist functions, constructing a self-representation of a people exceptionally close to nature, to and through the reindeer; this representation, utilizing the timely environmentalist discourses, had become attractive as conflicts up north had in one sense developed into disputes over the wisest use of northern nature. If for von Linné reindeer agency was of interest in its own right, and for Tanner it was to have an impact in the categorization of groups of people, for Sverloff the attached meanings of reindeer agency serve the function of re-coding the people by means of a reindeer seeking their vicinity.

Conclusions

When considering the gaze upon the Sami, it is easy to see a progression from hierarchical-mercantilist, via weakly anti-colonial to post-colonial. Also, as the timely discourses harden, human agency becomes more dominant and the scholarly agendas acquire more weight, the reindeer is given an increasingly instrumental role in discourses. von Linné's sense of awe gives way to Tanner's and Sverloff's way of utilizing the reindeer to show the placing of folk-groups in the hierarchies.

From the point of view of agency, in a semi-domesticated modus, reindeer can act in both spheres, between the unpredictable and autonomous natural state, beyond the reach of domestication, and the material, cultural world, dominated by human agency, where the predictable aspects of reindeer behavior can be harnessed for human use. The reindeer is an agent trafficking between nature and culture, borrowing from both realms, keeping the border between nature and culture wide open. In the traffic, reindeer are molded by human intrusion and returned to their natural environment in a transformed state, as both private property, more resistant to some threats in nature (through vaccination), thus changing its "balance," and as a commodity to be utilized as nutrition in mechanized human production.[48]

The significance of the reindeer is manifold. She carries a tremendous cultural and identificational significance for the Sami,[49] as well as molds social and economic fields as well as the fields of identity politics. In the socio-economic context, her dead corpse has the most significant effect. The bodily aspect is not present in the representational field, but enter the political. A third context, pure of human presence, best highlights the choice of the reindeer and her agency. As the southern gentleman-scholar makes his way back home, and the herder takes a break, the reindeer enters another context of her own choice: she checks to see whether the stock shows any signs of alarm, then starts looking for something to eat.

Notes

1 von Linné, *Lapinmatka 1732*; Tanner, "Antropogeografiska studier"; Sverloff, *Suenjelin saamelaisten perintö*.
2 Franklin, Lury and Stacey, "Introduction," 9.
3 Tønnessen, "Wolf Land," 289.
4 Aaltola, "Personhood and Animals," 175, 177; Weil, *Thinking Animals*, 139–40.
5 See Weil, *Thinking Animals*, 5–7.
6 Dransart, "Living Beings," 10; Das, "Being Together with Animals," 19, 28–9.
7 Ritvo, "Introduction," 9, 11.
8 Aaltola, "Personhood and Animals," passim; McFarland and Hediger, "Approaching the Agency," passim; Tønnesen, "Wolf Land," passim.
9 Tønnessen, "Wolf Land," 295.
10 Nance, *Entertaining Elephants*, 4–10, 97, 107, 175, 187, 195–8 and passim.
11 Nyyssönen and Salmi, "Towards a Multiangled Study," 42–4; von Uexküll and Joseph D. O'Neil, *A Foray into*; Weil, *Thinking Animals*, 31–2.
12 Compare HadžiMuhamedovi , "The Tree of Gernika," 57–8; compare Helminen, "Konikapinan Hilppa," 104 and passim.

13 Ritvo, *Animal Estate*, 5–6.
14 Ritvo, "Race, Breed," passim.
15 Niemelä, *Vain hyödynkö tähden?*, 122–36.
16 Mathisen, "Hegemonic Representations of Sámi Culture," 22.
17 von Linné charted the taxonomy of the insect in 1758. The plague is known throughout the caribou/reindeer herding world – the insect plants a parasite larva under the skin of the caribou/reindeer. ITIS, "ITIS report."
18 See also Turi, *Min bok om samene*, 59.
19 ATUM, AVT, Box 2, Folder 3, Om reinens ernæring. Carol. Linnaeus: Om renarnas Brömskulor i Lapland. Utdrag af Kungl. Svenska Vetenskaps Academiens Handlingar för år 1739; von Linné, *Lapinmatka 1732*, quote p. 71, 142.
20 Turi adds another way in which reindeer try to deal with the fly, by running around, out of fear, and climbing high cliffs to where it is cooler, with snow and glaciers. Turi, *Min bok om samene*, 38.
21 Ritvo, "Learning from Animals," 34.
22 For example, von Linné, *Lapinmatka 1732*, 31.
23 von Linné, *Lapinmatka 1732*, 113, 115.
24 Isaksson, *Kumma kuvajainen*, 46–7; von Linné, *Lapinmatka 1732*, 76, 79, 163.
25 von Linné, *Lapinmatka 1732*, 60, 139, 149. See also Ruotsala, *Muuttuvat palkiset*, 13. Other ways humans contribute to the reindeer are for example assisting the reindeer in digging for nutrition during wintertime (Heikkinen, *Sopeutumisen mallit*, 215) as well as guiding the stock to the forest in cases of storm. Johan Turi mentions herder assistance in cases of difficult calving. Turi, *Min bok om samene*, 80–1, 84.
26 Isaksson, *Kumma kuvajainen*, 57–62.
27 Dransart, "Living Beings," 10–1.
28 Ritvo, *Animal Estate*, 10–4.
29 Isaksson, *Kumma kuvajainen*, 87.
30 ATUM, AVT, Box1, Folder 1, Travel record, 26.8.1910, 30.8.1910.
31 Turi writes about the reindeer as a teacher, instructing the Sami in some aspects of the weather. Turi, *Min bok om samene*, 50.
32 ATUM, AVT, Box 6, Folder 3, Manus Petsamo, undated manuscript "Näringarnas förfall." The idea came from the Russian ethnographer Charuzin, which Tanner developed further.
33 Tanner, "Antropogeografiska studier," 154.
34 Gibson, *The Ecological Approach*, 36–7, 127–43; Warkentin, "Whale Agency," 23, 26–30.
35 See also Paulaharju, *Kolttain mailta*, 74–5.
36 Tanner, *Ihmismaantieteellisiä tutkimuksia*, 68. Author's translation.
37 Ruotsala, "Kaunista ja toimivaa," 361.
38 ATUM, AVT, Box 2, Folder 6c, notes: note entitled "Nielsen 1910, (Konrad…)"; Tanner, "Antropogeografiska studier," 18–21.
39 ATUM, AVT, Box 2, Folder 6c, notes: note titled "Nielsen 1910, (Konrad…)."
40 Naske and Slotnick, *Alaska*, 57, 68–9, 71–3.
41 Mathisen, "Hegemonic Representations," 23.
42 Tanner, "Antropogeografiska studier," 25–7.
43 Sverloff, *Suenjelin saamelaisten perintö*, 25, 55ff, 98, 101.
44 Sverloff, *Suenjelin saamelaisten perintö*, 67–8, 70–2.
45 Ritvo, "Learning from Animals," 39–46; Weil, *Thinking Animals*, 19.
46 Weil, *Thinking Animals*, 2012, 19–20, 83.
47 On the emotions of the reindeer, see Ryd, *Ren och varg*, 17–37, 41, 44, 57–8; Turi, *Min bok om samene*, 66–7, 82.
48 Compare Franklin, Lury and Stacey, "Introduction," 8–9; Ritvo, "Counting Sheep," 199–200; Ritvo, "How Wild is Wild?," 19.
49 Ruotsala, *Muuttuvat palkiset*, 91–107.

Bibliography

Archival Sources

Archive of Tromsø University Museum (ATUM), Norway, Archive of Väinö Tanner (AVT).
Box 1, Folder 1, Travel record, 26.8.1910, 30.8.1910.
Box 2, Folder 3, Om reinens ernæring.
Box 2, Folder 6c, notes: note entitled "Nielsen 1910, (Konrad…)."
Box 6, Folder 3, Manus Petsamo, undated manuscript "Näringarnas förfall."

Literature

Aaltola, Elisa. "Personhood and Animals." *Environmental Ethics* 30 (2008): 175–93.

Das, Veena. "Being Together with Animals: Death, Violence and Noncruelty in Hindu Imagination." In *Living Beings: Perspectives on Interspecies Engagements*, edited by Penelope Dransart, 17–32. London and New York: Bloomsbury, 2013.

Dransart, Penelope. "Living Beings and Vital Powers: An Introduction." In *Living Beings: Perspectives on Interspecies Engagements*, edited by Penelope Dransart, 1–16. London and New York: Bloomsbury, 2013.

Franklin, Sarah, Celia Lury, and Jackie Stacey. "Introduction." In *Global Nature, Global Culture*, edited by Sarah Franklin, Celia Lury and Jackie Stacey, 1–16. London: Sage Publications, 2000.

Gibson, James J. *The Ecological Approach to Visual Perception*. Hillsdale, NJ: Lawrence Erlbaum, 1986.

HadžiMuhamedovi , Safet. "The Tree of Gernika: Political Poetics of Rootedness and Belonging." In *Living Beings: Perspectives on Interspecies Engagements*, edited by Penelope Dransart, 53–72. London and New York, Bloomsbury, 2013.

Heikkinen, Hannu. *Sopeutumisen mallit, Poronhoidon adaptaatio jälkiteolliseen toimintaympäristöön Suomen läntisellä poronhoitoalueella 1980–2000*. Helsinki: SKS, 2002.

Helminen, Katri. "Konikapinan Hilppa ja pirtuhevonen Lapukka kerrottuina ja muistettuina toimijoina." In *Pappi, partisaani ja pirtuhevonen – poimintoja pohjoisen historiasta*, edited by Marianne Junila, Petri Granberg, Jukka Juntunen, Pia Kaitasuo, and Harri Turunen, 104–25. Oulu, Finland: Oulun Historiaseura, 2013.

Isaksson, Pekka. *Kumma kuvajainen, Rasismi rotututkimuksessa, rotuteorioiden saamelainen ja suomalainen fyysinen antropologia*. Pohjoiset historiat 1. Inari, Finland: Kustannus Puntsi, 2001.

ITIS. "ITIS report." Accessed August 29, 2013. www.itis.gov/servlet/SingleRpt/SingleRpt?search_topic=TSN&search_value=151808.

Linné, Carl von [Linnaeus]. *Lapinmatka 1732, Iter Lapponicum*. First edition 1889. Translated by Tuomo Itkonen and Ilkka Kukkonen. Hämeenlinna, Finland: Karisto, 1993.

Mathisen, Stein R. "Hegemonic Representations of Sámi Culture: From Narratives of Noble Savages to Discourses on Ecological Sámi." In *Creating Diversities, Folklore, Religion and Politics of Heritage*, edited by Anna-Leena Siikala, Barbro Klein and Stein R. Mathisen, 17–30. Helsinki: Finnish Literature Society, 2004.

McFarland, Sarah E., and Ryan Hediger. "Approaching the Agency of Other Animals." In *Animals and Agency: An Interdisciplinary Exploration*, edited by Sarah E. McFarland and Ryan Fediger, 1–20. Leiden, Netherlands: Brill, 2009.

Nance, Susan. *Entertaining Elephants, Animal Agency and the Business of the American Circus*. Baltimore, MD: The Johns Hopkins University Press, 2013.

Naske, Claus-M., and Herman E. Slotnick. *Alaska: A History of the 49th State*. Norman, OK: Oklahoma University Press, 1994.

Niemelä, Jari. *Vain hyödynkö tähden? Valistuksen ajan hyötyajattelun, luonnontieteen ja talouspolitiikan suhde Pehr Adrian Gaddin elämäntyön kautta tarkasteltuna*. Helsinki: Suomen Historiallinen Seura, 1998.

Nyyssönen, Jukka, and Anna-Kaisa Salmi. "Towards a Multiangled Study of Reindeer Agency: Overlapping Environments, and Human-Animal Relationships." *Arctic Anthropology* 50 (2013): 40–51.

Paulaharju, Samuli. *Kolttain mailta, Kansatieteellisiä kuvauksia Kuolan Lapista*. Helsinki: SKS, (1921) 2009.

Ritvo, Harriet. *Animal Estate: The English and Other Creatures in the Victorian Age*. London: Penguin Books, 1990.

Ritvo, Harriet. "Counting Sheep in the English Lake District: Rare Breeds, Local Knowledge, and Environmental History." In *Noble Cows and Hybrid Zebras: Essays on Animals and History*, 186–202. Charlottesville, VA and London: University of Virginia Press, 2010.

Ritvo, Harriet. "Introduction." In *Noble Cows and Hybrid Zebras: Essays on Animals and History*, 1–12. Charlottesville, VA and London: University of Virginia Press, 2010.

Ritvo, Harriet. "Learning from Animals: Natural History for Children in the Eighteenth and Nineteenth Centuries." In *Noble Cows and Hybrid Zebras: Essays on Animals and History*, 29–49. Charlottesville, VA and London: University of Virginia Press, 2010.

Ritvo, Harriet. "Race, Breed and Myths of Origin: Chillingham Cattle as Ancient Britons." In *Noble Cows and Hybrid Zebras: Essays on Animals and History*, 132–56. Charlottesville, VA and London: University of Virginia Press, 2010.

Ritvo, Harriet. "How Wild Is Wild?" In *The Edges of Environmental History: Honouring Jane Carruthers*, edited by Christof Mauch and Libby Robin, 19–24. Munich: Rachel Carson Center Perspectives, 2014.

Ruotsala, Helena. *Muuttuvat palkiset, Elo, työ ja ympäristö Kittilän Kyrön paliskunnassa ja Kuolan Luujärven poronhoitokollektiiveissa vuosina 1930–1995*. Helsinki: Suomen Muinaismuistoyhdistys, 2002.

Ruotsala, Helena. "Kaunista ja toimivaa – saamelaisten aineellinen kulttuuri." In *Saamentutkimus tänään*, edited by Irja Seurujärvi-Kari, Petri Halinen and Risto Pulkkinen, 344–68. Helsinki: SKS, 2011.

Ryd, Yngve. *Ren och varg, Samer berättar*. Stockholm: Natur och kultur, 2007.

Sverloff, Matti. *Suenjelin saamelaisten perintö*. S.p., Matti Sverloff and Maa- ja metsätalousministeriö, 2003.

Tanner, Väinö. "Antropogeografiska studier inom Petsamo-området. 1 Skoltlapparna." *Fennia* 49 (1929): 1–518.

Tanner, Väinö. *Ihmismaantieteellisiä tutkimuksia Petsamon seudulta 1, Kolttalappalaiset*, edited by Paulo Susiluoto. Helsinki: SKS, 2000.

Turi, Johan. *Min bok om samene*. Karasjok, Norway: ČálliidLágádus, 2011.

Tønnessen, Morten. "Wolf Land." *Biosemiotics* 3 (2010): 289–97.

Uexküll, Jakob von, and Joseph D. O'Neil. *A Foray into the Worlds of Animals and Humans: With a Theory of Meaning*. Minneapolis, MN and London: University of Minnesota Press, 2010.

Warkentin, Traci. "Whale Agency: Affordances and Acts of Resistance in Captive Environments." In *Animals and Agency: An Interdisciplinary Exploration*, edited by Sarah E. McFarland, and Ryan Fediger 23–43. Leiden: Brill, 2009.

Weil, Kari. *Thinking Animals, Why Animal Studies Now?* New York: Columbia University Press, 2012.

11 Anthropogenic Food Sources in the Co-Existence of Humans with Liminal Animals in Northern Environments

Representations from *National Geographic* Magazine

Linda Kalof, Cameron Whitley, Stephen Vrla, and Jessica Bell Rizzolo

Introduction

Our chapter examines the visual representation of liminal animals over the past 126 years in *National Geographic*, a widely read popular science magazine. Liminal animals are non-domesticated species who live among humans and are neither fully members of the human community, nor fully external to it: animals such as birds, raccoons, coyotes, squirrels, and rats.[1] We follow O'Connor's argument that the complex and fluid interrelationships between humans and liminal, or commensal, animals form a vital community that is sustained by trophic relations and offers unique opportunities for human contact with other animals. As nature becomes ever more commoditized, the animals who live independently among us will assume a greater importance in giving people a connection to species other than domestic companions and livestock.[2]

We examine the representation of liminal animals in northern environments. Northern environments are unique in that they have geographical and climatic challenges related to cold and inaccessibility, thus creating a "valuable analytical framework that … transgresses borders" allowing a better understanding of the shifting human–animal relationships in urban, rural, and wilderness contexts.[3] As scholars have recently noted, nature and culture are "entangled with each other" in northern environments, generating a special need for historical research that engages with interdisciplinarity.[4] For example, in a historical study, Finstad documented how vermin – rats, lice, and bedbugs – contributed to the shaping of the urban community and public health in Trondheim, Norway, in the first half of the twentieth century.[5] O'Connor has also focused on the shifting relationships of humans and liminal animals with an emphasis on northern Europe. He notes the importance of northern environments for some liminal species, such as cats who, in higher latitudes, have a marked adaptation to scavenging a wide variety of foods from human garbage.[6] In addition, he argues that the decline in some urban bird populations is related to rising affluence: bird populations remain stable in poorer neighborhoods, but decline in areas with two-car households and garden makeovers that include patios "for burgers and Chardonnay."[7]

Our examination of *National Geographic* animal photography emerges from (1) the argument that the print media configures cultural understandings of the relationship between humans and other animals,[8] and (2) the documentation that animal images reflect cultural messages and historical shifts.[9] *National Geographic* is unique among print media in that it has shaped popular ideas of people, places, animals, and the natural world for 126 years. Considered the gold standard for the representation of the natural world,[10] *National Geographic* is published in 34 languages, has a global monthly readership of 8.2 million,[11] and is renowned for its glossy photographs.[12] The twentieth century was a time of sweeping re-evaluation of the relationship between humans and other animals driven primarily by the increased scientific and popular awareness of environmental destruction and the dramatic loss of biodiversity and animal habitat. Indeed, over the last half century, historians and sociologists of science have become increasingly aware of how popular culture influences the science and politics surrounding the affiliation between humans and other animals, and a study of the popular science visual representations of the important relationship between humans and their liminal animal neighbors is long overdue. Our research fills that gap.

Background

Liminal Animals in Northern Environments

Historically, many animals in northern environments have relied upon, and been influenced by, anthropogenic food sources. In Denmark, biologists have documented changes in skull sizes in badger and red fox populations, which they attribute to changes in human-provided food sources.[13] In Finland, red foxes have adjusted to urban living by increasingly relying on anthropogenic food sources, habituating to the presence of humans, and using human-made structures for shelter.[14] In Russia, urban common hamsters now demonstrate features not seen in non-urban populations, such as "adaptations of diet to supplemental food resources, tolerance toward anthropogenic structures and genetic modifications"[15] and feral dogs have become dependent upon human food sources.[16] In Alaska, bald eagles and moose tend to congregate or increase in population near urban centers likely to have access to food and shelter sources.[17]

Reactions toward liminal animals in northern environments vary depending upon the individual species and environmental contexts. Songbirds in Finland have been encouraged through feeding and constructing nest boxes,[18] and in Anchorage, Alaska, moose are considered a "symbol of the city" and "help engender a sense of place."[19] However, human communities can sometimes become intolerant of urban wildlife, particularly ungulates and the browsing, disease transmission and vehicle collisions to which they contribute.[20]

Finally, it is important to discuss how the unique geographical and cultural characteristics of northern environments may inform or alter the treatment of liminal animals. In Scandinavia, the forest is an important part of inhabitants' sense of place, and there is a long tradition of interacting with forests through hunting and fishing.[21] The circumpolar regions of the Arctic are also marked by unique characteristics such as a sparse population distributed over a vast geographic area, seasonal extremes of temperature and daylight, a rapidly changing climate that is very sensitive to global warming, and the presence of indigenous groups such as the Inuit who hold distinct notions about human–environment relationships.[22] For example, the subsistence and land claims of Alaskan natives often conflict with the worldviews of European-Americans, leading Alaska to become a landscape with many contesting senses of place,[23] and contrasting senses of place affect which interactions with animals are considered legitimate, as animals who transgress spatial and social boundaries are often labeled as "problem animals" or pests.

Liminal Animals in the Media

Several scholars have examined how written media contribute to social representations of liminal animals. In their content analysis of newspaper articles about coyotes, Alexander and Quinn found incongruence between the perceived threat of coyotes and the actual number of coyote attacks. They postulated that, since many urban residents viewed the coyote as unnatural or out of place in cities, even benign sightings of coyotes in urban areas became risky "incidents."[24] The finding that amplification of risk can occur when an animal is viewed as out of place is also supported by content analyses of newspaper articles on other animals who live near or in urban areas, such as cougars[25] and pigeons.[26] Through selecting and circulating particular framings of liminal animals, the media influences cultural understandings of them. For example, Jerolmack demonstrated how written media played a vital role in the expansion and acceptance of the negative representation of pigeons as "rats with wings."[27]

Visual media are a source of popular culture that has rarely been the focus of scholarly work on the representation of liminal animals. The exceptions are three books that focus on species who often live liminal lives: the rat, the pigeon, and the fox.[28] Artistic representations of the pigeon usually depict the animal as dove-like with associations to innocence and purity.[29] Moving film representations of the rat cluster the animal into two major themes: objects of horror and in "association with low life … [lending himself] to horror films either attacking humans or figuring as a totemic symbol of evil."[30]

The negative visual representation of the rat corroborates Malamud's argument that when animals are pictured in visual culture, the representations are always biased and serve the interests of humans; visual representations place animals in cultural constructions, hiding who the animals actually are – their visibility diminishes their authenticity. Malamud suggests asking a key ethical question when analyzing a visual representation: does the construction do more good than harm?[31] We take up Malamud's challenge by examining the representation of liminal animals in *National Geographic* photographs. Our research question is: How has popular science visually represented the relationship between humans and their liminal animal neighbors over the last 100 years?

Method

Our unit of analysis was the image and the associated caption of all representations of liminal animals in northern environments published over the past 126 years in a random sample of one issue per year of *National Geographic* magazine.[32] We also included in the sample the liminal animal representations uncovered in our ongoing larger project of collaborative research on *National Geographic*. We chose the 40th parallel North to operationalize "northern environment," and our focus was on animals who use human habitats for living space and food.

Agency and Animal Liminality

Agency is an important characteristic of an animal's liminality in the activities of feeding, building nests, rummaging in garbage dumps, and begging for food. We searched for representations of animals who were engaged in some kind of agentic behavior rather than represented as an object, for example constrained in a laboratory setting. We expected most of the agency exhibited by the animals would be based on their food search activities; O'Connor argues that access to food drives most commensal relationships because the "trophic relations of people and other animals are central to sustaining their relationship."[33]

Results

A variety of liminal species were represented in northern environments in the magazine, all of whom were depicted with some degree of agency in adapting to human food and habitat sources. There were 55 photographs of birds and 17 photographs of mammals, including bears, bighorn sheep, deer, moose, rats, and wolves. *National Geographic*'s depiction of our animal neighbors fell into three major themes: bird species encouraged by humans to join the human landscape with offerings of food and/or shelter, small animal-occupants who on their own utilize human food and/or space, and charismatic megafauna lured by food into the human landscape.

Bird Species Encouraged by Humans to Join the Human Landscape

The largest proportion of liminal animal photographs in the sample were of wild birds (76%), many of which were from the early years of *National Geographic*. These images often depicted humans encouraging the presence of birds both by building shelters such as birdhouses and, in particular, by offering food. It is no surprise that the liminal animals in the early years were often birds, giving homage to "birdwatching," an activity established in the early eighteenth century. The *National Geographic* bird photographs also reflect the deep symbolic place birds have occupied in human history. In the Classical world, statues illustrate birds as signs of godliness or perfection in childhood, with children holding birds in their hands and adults offering birds to children.[34]

Multiple photographs of birds illustrated a single article, particularly in the early years of our sample. When color found its way into *National Geographic* photographs, a 1947 article published 13 pictures of hummingbirds interacting with bird feeders and humans. A number of articles focused on humans' co-living with wild birds by providing food and shelter. Humans provided food for birds in a variety of ways, including erecting window boxes and food bells that automatically supply grain from a receptacle, fastening food to trees and, in two photographs, even offering food to birds by mouth: "On intimate terms: This jolly little white-breasted nuthatch has just taken a dainty morsel from the lips of its friend."[35]

Humans' interest in being on close terms with wild birds was well articulated in a 1914 article "Encouraging Birds around the Home" that discussed "companionable birds" with surprising empathy, such as chickadees whose "tameness, quaint notes, and friendly ways make them general favourites"[36] and bluejays:

> Bluejays feeding in a weathercock food-house: The same man builds also a sheltered food-house that turns with the wind like a weather vane, so as to present always a lee side for the better protection of the birds.[37]

Notably, some captions addressed the uniqueness of northern environments as a factor in the reason to offer birds food. For example, a photograph of a covey of quail feeding on a snow-covered field had this caption:

> Flock of quail at a food station: In bad weather, particularly in the North, where we are so apt to be covered up with snow, more artificial means of feeding should be resorted to, and food stations, food-houses, and food shelters of various sorts should be established in proper places.[38]

Figure 11.1 Providing food for quail in a snow-covered field.

Source: Photographer Ernest Harold Baynes/National Geographic Creative, Image 1314533, published March 1914, 329.

The photographs also depicted humans attracting birds by building birdhouses and birdbaths. The houses ranged from rough pieces of log to elaborate two-story structures, boxes positioned close to house windows, and residences that provide both shelter and food. The provision of some kind of bathing facility was considered essential in attracting wild birds.[39]

These analyses of visual narratives in *National Geographic* provide insight into the bird–human relationship, including the development and evolution of birdwatching as a sociocultural trend and our evolving relationship with liminal animals more broadly.

Small Animal-Occupants Who on Their Own Utilize Human Food and/or Space

Liminal animals who utilize human food and space were represented in 18 percent of the sample. These photographs symbolize liminal animals as humans most commonly see them: as "pests," "vermin," or the uninvited. Most of the images were of animals, particularly birds, peacefully resting, foraging, or passing through urban areas, such as waterfowl swimming in a park in Helsinki[40] and ducks swimming in a partially ice-covered lake in Stockholm.[41] The activity level of these animals varied in the representations. Canada geese were photographed flying across the Manhattan skyline,[42] and a stunning photograph of a flock of mallard ducks illustrated their flapping frenzy on liftoff from a pond in South Dakota when menaced by hawks overhead.[43]

Similar to our first theme, the most frequent representations of animal-occupants were birds who take up residence in the human-built environment, particularly barns.[44] While the uninvited may be visually appealing for a time, they may easily become a problem. Humans' enduringly ambivalent attitudes toward liminal animal neighbors were well illustrated by photographs of animal-occupants who become animal-invaders, such as panhandling ungulates.[45]

Our complex relationship with liminal animals evolves over time. The twenty-first century has ushered in a new perspective on animal-occupants turned invaders into the human landscape. For instance, animal-invaders were once persecuted, such as the rats of Copenhagen who were relentlessly exterminated by adults and children alike:

> Officials receiving rats killed by children during a campaign against them in Copenhagen: By a small reward to the juvenile members of the family for rats captured, the pests may be kept down and the primitive joys of the chase experienced by the young trappers.[46]

While the rat continues to be the object of fear and loathing among many humans, some species have effectively transitioned from pest to awe-inspiring, or maintain a complicated status. For example, one of the most successful animals at taking advantage of human food sources is the corvid, a highly intelligent omnivore. An article in the January 1999 issue described the transition of corvids from persecuted pests early in the twentieth century to animals on the rebound by taking advantage of anthropogenic food sources. Human regard for the corvid's intelligence and cunning is well illustrated in the following caption of a Yellowstone raven perched on a snowmobile extracting food from the vehicle's storage compartment:

> Street-smart in the wilderness, a hungry opportunist opens a snowmobile's storage compartment, making quick work of its Velcro-fastened flap. It extracted a crumpled map in its hunt for such favoured items as cookies and potato chips. "Ravens adapt quickly to changing technology" ... They can untie knots and unzip zippers, making off with car keys and other shiny objects ... [one must] secure food and valuables with straps and strong buckles.[47]

Nevertheless, the corvid is still considered by some an unwelcome invader of human space.[48]

Representations of animals who invade the human landscape are important indicators of the blurred nature of our constructions of particular species. For example, the geese so lovingly fed by a woman in a Toronto park[49] were pictured as bothersome pests in a photograph of a gaggle of Canada geese looking for food on a golf course as human golfers stepped around the animals to continue their game.[50]

Some photographs highlighted the challenge of managing invading species, such as the urban raccoon struggling to escape a live trap:

Figure 11.2 Non-lethal management of rabies among urban raccoons.

Source: Photographer Karen Kasmauski/National Geographic Creative, Image 485366, published July 1994, 78.

> Border defenses: furious at being confined, an urban raccoon faces a shot of rabies vaccine in Ontario, Canada, during a campaign to shut out an epidemic now creeping up the eastern US.[51]

The representation of uninvited liminal animals in *National Geographic* clearly depicts the complicated and often controversial place that these animals occupy in human culture. In some images they are admired, while in others they are represented as a nuisance. Increasingly, scholars have been interested in sociocultural depictions of the uninvited, paying particular attention to how we create urban spaces that restrict or invite non-human guests. What is evident from these images is not only the "pest" narrative, but also a more recent undercurrent to establish and construct spaces where humans and animals may co-exist.

Charismatic Megafauna Lured by food into the Human Landscape

Whereas providing foods, water, and shelter for birds has few negative consequences for the humans or birds involved, feeding megafauna can lead to property damage, human injury, and, critically, the displacement of the animal or her eventual death. This theme fit nearly eight percent of our sample. Here we encounter a human–animal interactive strategy popular since antiquity – using food to bring wild animals close to humans. For example, Varro described a country estate whose owner wanted animals readily available to kill. To achieve this feat, the staff trained wild animals to appear at the sound of a trumpet – played by a musician dressed as Orpheus – by placing food in the same location every day.[52] While some motivations for luring megafauna may be different today than they were in antiquity, the practice remains resilient, and we found that food was used to attract megafauna often for the protection of individual animals or as a form of wild animal tourism.

One might assume that the desire to assist an orphaned animal would be more evident in the later years of *National Geographic*, a time more closely aligned with the beginning of the environmental and conservation movements. However, we found that images of humans assisting orphaned animals spanned the entire time period of the sample, a not surprising finding given the magazine's focus on conservation:

> This two-weeks-old baby moose thrives on a nursing bottle: Not often is a moose seen this close to an automobile tag in Northern Minnesota. The ungainly looking youngster, when only a few hours old, was separated from its mother by a forest fire. It was rescued at Beaver Bay, about 50 miles from Duluth.[53]

Several photographs portrayed animals who had been intentionally tamed through feeding on a regular basis:

> For years Carl C. Murty, Jr., of Three Lakes (Wisconsin) has tamed wild foxes, otters, raccoons, and bears. These orphaned fawns visit him by night to beg food and to play with his dogs. Each drinks three quarters of milk daily. Red collar warns hunters of a pet.[54]

Figure 11.3 Deer tamed through regular feeding.

Source: Photographer Volkmar K. Wentzel/National Geographic Creative/Image 1106483, published February 1957, 174.

However, feeding deer can be particularly problematic; they are invaders of human space and are often called "rats with hooves" because of their penchant for eating the vegetation around human residences:

> Cloaked in evening shadows, a white-tailed buck forages the property of Mary Aponick's Philadelphia home. Ravenous deer forced her to build a seven-foot-high fence around her backyard vegetable garden … many suburban homeowners, farms, and timber operators want the population culled.[55]

Photographs of human spectators observing invited mammals also appear throughout the magazine's history. A 1949 photograph of a man feeding a mother bear is critical of this risky form of tourism, again suggesting the magazine's enduring focus on conservation:

> If this man stepped between mother and cub, he might be seriously injured: Even feeding her is dangerous. Back of these cars a score of more will pile up as drivers abandon them in the middle of the highway to leap to the scene, camera in hand. Most bear-tourist relations in Yellowstone National Park are cordial, yet such reckless tactics as these, in violation of National Park Service warnings, often result in clawing and bitings. Two persons have even been killed.[56]

Still, the awe humans have with getting close to and interacting with wildlife by feeding them continues unabated throughout the time period. In a photograph published in 1980, a bighorn sheep stands beside a vehicle stopped on the Trans-Canada highway. The sheep has his head inside the passenger side window looking for food, and the human occupants of the car are thrilled with the experience.[57]

Charismatic megafauna who enter the human landscape looking for food have become a pernicious problem of human–animal conflict made worse with the destruction of wildlife habitats. The danger of habituated wildlife and the contemporary strategy of dealing with problem animal-invaders was illustrated in an article on the polar bears of Churchill, Manitoba.[58] These iconic megafauna get jail time instead of slaughter as a way of controlling their invasion into human space. The town of Churchill has a problem with the local polar bears who, because of global warming, are stuck on the mainland until the Hudson Bay freezes over and they are able to migrate onto the frozen ice to hunt ringed seals. As a way of keeping the bears out of trouble during October and November, a time when they outnumber Churchill's human population of 800, a polar bear jail was erected in a former aircraft storage hangar. Problem bears are tranquilized for incarceration until the bay freezes over and they are removed by helicopter back into the wild. The presence of the bears is a tourist attraction, and tourists book Churchill's four hotels far in advance for the Polar Bear Alert during October and November. The bears were represented as attracted to the presence of humans and their food, such as the photograph of a bear on her hind legs peering into a tour vehicle following the aroma of frying bacon.[59]

The devastating consequences of such habituated behavior, however, were also illustrated in the article. For example, a dirty, soot-covered polar bear was photographed rummaging in a fiery garbage dump with the number 13 painted on his side:

> Waiting for the bay to freeze, bears invade the town and scavenge its nearby dump. Biologists use dye to identify chronically aggressive bears; three to five repeat offenders must be destroyed every year … Lucky Number 13, a subadult male, avoided the dump last season.[60]

We close with another compelling example of a carnivore habituated to human food sources, a co-living strategy that almost always spells death for the animal. A wolf was photographed in semi-darkness staring at the photographer from a close distance:

> Hunter-turned beggar, a wolf accustomed to handouts on Vargas Island stares down the photographer Joel Sartore hoping in vain for a scrap. One of the pack later attacked a camper. Two wolves were shot as a result.[61]

This photograph highlights the crucial, unfortunate difference between the feeding of birds and the feeding of megafauna in northern environments. In both

Figure 11.4 Habituated bear searching for food in a garbage dump.

Source: Photographer David Hiser/National Geographic Creative, Image 278049, published March 1982, 395.

cases, the animals entertain us with their appearances and behaviors. However, while the birds may need our offerings of food, water, and shelter to survive in harsh, frigid environments, such offerings do megafauna more harm than good.

Conclusion

Our work establishes that the *National Geographic* representations of human–liminal animal interactions over the last 100 years depict a multi-species community centered around anthropogenic food sources. The photographs of our animal neighbors fell into three major trophic-based themes: bird species encouraged by humans to join the human landscape by providing food and/or shelter, small animal-occupants who on their own utilize human food and/or space, and charismatic megafauna lured by food into the human landscape.

The majority of the liminal animal representations were of birds actively encouraged to live close to humans with the offer of food and/or shelter. The depictions of wild birds in the early years of *National Geographic* corroborates research documenting that birds have long been particularly well liked in northern environments, encouraged to join the landscape and fed for example by almost three-fourths of the contemporary population of Norway at the turn of the twentieth century.[62] Throughout the century, wildlife who occupy human space and/or scavenge food as opportunists were well represented. The blurred nature of some species such as geese was evident in photographs, some picturing the animal as aesthetically pleasing and others picturing her as bothersome in the human landscape. Some of the blurring had to do with the activity of the animal in the representation, with the pleasing animal waiting patiently for a food handout and the pest foraging for food a bit too close to human leisure activities.

Animal agency was well-illustrated in the representations of liminal animals in the northern environment. Birds consumed suet and seed, built nests, and bathed; ducks and geese swam on urban ponds, flew through the sky, and searched for food on a golf course; deer foraged backyard gardens, begged for food, and played with domestic dogs; a raven extracted food from a Velcro-fastened storage compartment; polar bears rummaged in garbage dumps and panhandled for food; and sadly a "wild" wolf was photographed begging for a handout. The one overwhelming variable in the majority of the representations of animal agency is their search for anthropogenic food sources.

Living in the northern environment was mentioned in the *National Geographic* narratives as a reason for providing food stations for birds who would otherwise not easily sustain themselves. But perhaps the most compelling examples of how living in the North compromised the search for a customary food source were the scavenging polar bears of Churchill, Manitoba. Because of global warming which brings a late freeze to the Hudson Bay, the bears cannot hunt seals and thus rely on garbage dumps and handouts for sustenance. Food handouts for wildlife are particularly problematic human–animal interactions,[63] and the human penchant for feeding wildlife has catastrophic consequences. In particular, human–carnivore interactions often lead to conflict in which the animal ends up the loser. *National Geographic* has appropriately emphasized this danger in most of the articles and captions of humans feeding wildlife published in the last half century, consistent with the increasing concern over the "necessary" destruction of wildlife who become problems in their search for anthropogenic food sources. Thus, *National Geographic*'s representations were beneficial to the animals themselves by emphasizing the dangerous context of habituation for some animal species, particularly large carnivores.

The positive representation of animals held throughout our sample. Only in the very early years of *National Geographic* were liminal animals depicted as loathsome pests such as the Copenhagen rats. While negative representations of liminal animals usually emerge from their status as boundary crossers, it is widely accepted in the animal studies community that we humans must find a way to co-live with wild species.[64] However, it is possible that the overall favorable imagery of liminal animals is the outcome of selecting representations in the northern environment for this study. Southern climes are noted for their pest species, including cockroaches and year-long active mosquitoes, who are not usually depicted in a positive light. While an intriguing hypothesis, a systematic examination of the difference in liminal animal representations by southern and northern geographic regions must wait for future research.

We agree with the argument that the future of thousands of species and many ecosystems depends on how humans interact with animals in the "urban jungle" – particularly those pigeons, raccoons, squirrels, and crows who are so abundant in our neighborhoods – and people are more likely to conserve nature when they have direct experiences with the natural world.[65]

Our work confirms that the visual representation of animals in *National Geographic* satisfies Malamud's ethical question – the constructions do more good than harm. As a print media exemplar of the popular science community, *National Geographic* is contributing to the rise of an ecological sensibility, and, as suggested by other scholars,[66] future environmental sustainability and animal conservation may be in the hands of people who live alongside and have regular interaction with liminal animals.

Acknowledgements

This research was supported by a grant from the National Science Foundation (Award SES-1247824).

Notes

1 Donaldson and Kymlicka, *Zoopolis*.
2 O'Connor, *Animals as Neighbors*.
3 Jørgensen and Sörlin, *Northscapes*, 4.
4 Jørgensen et al., "Entangled Environments," 9–34.
5 Finstad, "Rat Nest."
6 O'Connor, *Animals as Neighbors*, 58.
7 O'Connor, *Animals as Neighbors*, 106.
8 See Kalof, *Looking at Animals*; Malamud, *An Introduction to Animals*.
9 Arluke and Bogdan, *Beauty and the Beast*; Kalof and Amthor, "Cultural Representations of Problem Animals"; Dunaway, *Natural Visions*.
10 Montgomery, "Through a Lens."
11 *National Geographic*, "About the National Geographic Society."
12 Lutz and Collins, *Reading National Geographic*.
13 Yom-Tov, Yom-Tov, and Baagøe, "Increase of Skull Size."
14 Vuorisalo et al., "Urban Red Foxes."
15 Feoktistova et al., "The Common Hamster," 115.
16 Ivanter and Sedova, "Ecological Monitoring of Urban Groups."
17 Johnson, "Nesting Bald Eagles."
18 Vuorisalo et al., "Urban Biodiversity in Local Newspapers."
19 Whittaker et al., "Understanding Beliefs and Attitudes," 1115.
20 Ibid.
21 Rydberg and Falck, "Urban Forestry in Sweden."
22 Parkinson, "The International Polar Year."
23 Behnke, "Alaska's Contested Rural Landscapes."
24 Alexander and Quinn, "Coyote (Canis latrans) Interactions."
25 Wolch, Gullo, and Lassiter, "Changing Attitudes."
26 Jerolmack, "How Pigeons Became Rats."
27 Jerolmack, "How Pigeons Became Rats."
28 Burt, *Rat*; Allen, *Pigeon*; Wallen, *Fox*.
29 Allen, *Pigeon*, 145, 155, 158.
30 Burt, *Rat*, 85.
31 Malamud, *An Introduction to Animals*.

32 The random sample of 2243 images was collected as part of a National Science Foundation funded project and stored as Endnote citation records.
33 O'Connor, *Animals as Neighbors*, 7.
34 Kalof, *Looking at Animals*, 22–3.
35 *National Geographic*, March 1914, 333.
36 *National Geographic*, March 1914, 330.
37 *National Geographic*, March 1914, 326.
38 *National Geographic*, March 1914, 329. See Figure 11.1.
39 *National Geographic*, March 1914, 340.
40 *National Geographic*, May 1925, 611.
41 *National Geographic*, October 1928, 408.
42 *National Geographic*, February 1992, 66–7.
43 *National Geographic*, October 2007, front matter.
44 *National Geographic*, March 1914, 324.
45 *National Geographic*, September 1966, 371.
46 *National Geographic*, July 1917, 13.
47 *National Geographic*, January 1999, 112–3.
48 For an exception of this view, see Heinrich, *Mind of the Raven*.
49 *National Geographic*, August 1973, 178–9.
50 *National Geographic*, February 1992, 82–3.
51 *National Geographic*, July 1994, 78. See Figure 11.2.
52 Shelton, "Beastly Spectacles," 119.
53 *National Geographic*, March 1935, 318.
54 *National Geographic*, February 1957, 174. See Figure 11.3.
55 *National Geographic*, February 1992, 72.
56 *National Geographic*, June 1949, 739.
57 *National Geographic*, June 1980, 764–5.
58 *National Geographic*, March 1982.
59 *National Geographic*, March 1982, 394–5.
60 *National Geographic*, March 1982, 395. See Figure 11.4.
61 *National Geographic*, February 2003, 124–5.
62 See Vuorisalo et al., "Urban Red Foxes"; Bjerke and Østdahl, "Animal-Related Attitudes and Activities."
63 Breck, Lance, and Seher, "Selective Foraging."
64 See DeStefano, *Coyote at the Kitchen Door*.
65 Dunn et al., "The Pigeon Paradox."
66 Dunn et al., "The Pigeon Paradox"; O'Connor, *Animals as Neighbors*.

Bibliography

Alexander, Shelley M., and Michael S. Quinn, "Coyote (Canis latrans) Interactions with Humans and Pets Reported in the Canadian Print Media (1995–2010)." *Human Dimensions of Wildlife* 16 (2011): 345–59.
Allen, Barbara. *Pigeon*. London: Reaktion Books, 2009.
Arluke, Arnold, and Robert Bogdan. *Beauty and the Beast: Human-Animal Relations as Revealed in Real Photo Postcards, 1905–1935*. Syracuse, NY: Syracuse University Press, 2010.
Behnke, Steven. "Alaska's Contested Rural Landscapes and the Subsistence Claims of Alaska Natives." In *Geographical Identities of Ethnic America: Race, Space, and Place*, edited by Kate A. Berry and Martha L. Henderson, 149–73. Reno, NV: University of Nevada Press, 2002.
Bjerke, Tore, and Torbjørn Østdahl. "Animal-Related Attitudes and Activities in an Urban Population." *Anthrozoös* 17 (2004):109–29.

Breck, Stewart. W., Nathan Lance, and Victoria Seher. "Selective Foraging for Anthropogenic Resources by Black Bears: Minivans in Yosemite National Park." *Journal of Mammalogy* 90 (2009): 1041–44.

Burt, Jonathan. *Rat*. London: Reaktion Books, 2006.

DeStefano, Stephen. *Coyote at the Kitchen Door: Living with Wildlife in Suburbia*. Cambridge, MA: Harvard University Press, 2010.

Donaldson, Sue, and Will Kymlicka. *Zoopolis: A Political Theory of Animal Rights*. New York: Oxford University Press, 2011.

Dunaway, Finis. *Natural Visions: The Power of Images in American Environmental Reform*. Chicago: University of Chicago Press, 2008.

Dunn, Robert R., Michael C. Gavin, Monica C. Sanchez, and Jennifer N. Solomon. "The Pigeon Paradox: Dependence of Global Conservation on Urban Nature." *Conservation Biology* 20 (2006): 1814–6.

Feoktistova, Natalia Yu, Alexey Surov, Nikolay N. Tovpinetz, M. V. Kropotkina, Pavel L. Bogomolov, Carina Siutz, Werner Haberl, and Ilse E. Hoffmann. "The Common Hamster as a Synurbist: A History of Settlement in European Cities." *Zoologica Poloniae* 58 (2013): 116–29.

Finstad, Terje. "Rat Nest: Stories about Bugs in Trondheim in the First Half of the 1900s." *Historisk Tidsskrift* 92 (2013): 35–61.

Heinrich, Bernd. *Mind of the Raven*. New York: Harper Collins, 2009.

Ivanter, E. V., and N. A. Sedova. "Ecological Monitoring of Urban Groups of Stray Dogs: An Example of the City of Petrozavodsk." *Russian Journal of Ecology* 39 (2008): 105–10.

Jerolmack, Colin. "How Pigeons Became Rats: The Cultural-Spatial Logic of Problem Animals." *Social Problems* 55 (2008): 72–94.

Johnson, Nathan P. "Nesting Bald Eagles in Urban Areas of Southeast Alaska." In *Bald Eagles in Alaska*, edited by Bruce A. Wright and Phil Schempf, 325–43. Juneau, AK: Bald Eagle Research Institute, University of Alaska Southeast, 2008. Accessed October 8, 2014. http://raptors.hancockwildlife.org/BEIA/PAGES/Section-37.pdf.

Jørgensen, Dolly, and Sverker Sörlin. *Northscapes: History, Technology and the Making of Northern Environments*. Vancouver: UBC Press, 2013.

Jørgensen, Finn Arne, Unnur Birna Karlsdóttir, Erland Mårald, Bo Poulsen, and Tuomas Räsänen. "Entangled Environments: Historians and Nature in the Nordic Countries." *Historisk Tidsskrift* 92 (2013): 9–34.

Kalof, Linda. *Looking at Animals in Human History*. London: Reaktion, 2007.

Kalof, Linda, and Ramona Fruja Amthor. "Cultural Representations of Problem Animals in *National Geographic*." *Etudes Rurales* 185 (2010): 165–80.

Lutz, Catherine A., and Jane L. Collins. *Reading National Geographic*. Chicago: University of Chicago Press, 1993.

Malamud, Randy. *An Introduction to Animals and Visual Culture*. New York: Palgrave, 2012.

Montgomery, Scott. "Through a Lens, Brightly: The World According to *National Geographic*." *Science as Culture* 4 (1993): 7–46.

National Geographic, 1888–2014.

National Geographic. "About the National Geographic Society." Accessed November 1, 2012. http://press.nationalgeographic.com/about-national-geographic.

O'Connor, Terry. *Animals as Neighbors: The Past and Present of Commensal Species*. Lansing, MI: Michigan State University Press, 2013.

Parkinson, A. J. "The International Polar Year, 2007–2008. An Opportunity to Focus on Infectious Diseases in Arctic Regions." *Emerging Infectious Diseases* 14 (2008): 1–3.

Rydberg, D., and J. Falck. "Urban Forestry in Sweden from a Silvicultural Perspective: A Review." *Landscape and Urban Planning* 47 (2000): 1–18.

Shelton, Jo-Ann. "Beastly Spectacles in the Ancient Mediterranean World." In *A Cultural History of Animals in Antiquity*, edited by Linda Kalof, 97–126. Oxford, UK: Berg, 2007.

Vuorisalo, Timo, Rauno Lahtinen, and Hannu Laaksonen. "Urban Biodiversity in Local Newspapers: A Historical Perspective." *Biodiversity & Conservation* 10 (2001): 1739–56.

Vuorisalo, Timo, Kati Talvitie, Kaarina Kauhala, Auli Bläuer, and Rauno Lahtinen. "Urban Red Foxes (*Vulpes vulpes* L.) in Finland: A Historical Perspective." *Landscape and Urban Planning* 124 (2014): 109–17.

Wallen, Martin. *Fox*. London: Reaktion Books, 2006.

Whittaker, Doug, Michael J. Manfredo, Peter J. Fix, Rick Sinnott, SuzAnne Miller, and Jerry J. Vaske. "Understanding Beliefs and Attitudes about an Urban Wildlife Hunt near Anchorage, Alaska." *Wildlife Society Bulletin* 29 (2001): 1114–24.

Wolch, Jennifer R., Andrea Gullo, and Unna Lassiter. "Changing Attitudes toward California's Cougars." *Society and Animals* 5 (1997): 95–116.

Yom-Tov, Y., S. Yom-Tov, and H. Baagøe. "Increase of Skull Size in the Red Fox (*Vulpes vulpes*) and Eurasian Badger (*Meles meles*) in Denmark during the Twentieth Century: An Effect of Improved Diet?" *Evolutionary Ecology Research* 5 (2003): 1037–48.

12 Changing Narratives of Human

Large Carnivore Encounters in Nineteenth-Century Sweden

Karin Dirke

Introduction

The frameworks within which human–large carnivore encounters were narrated in Sweden changed during the nineteenth century. As the practises of hunting, zoology and the discourse about such matters changed, the view of the animals differed. This chapter aims at discussing this development as it was expressed in hunting stories, zoology and newspapers. Since the discourse on hunting during the nineteenth century primarily concerned the brown bear (*Ursus arctos*), and to some extent the gray wolf (*Canis lupus*), these are the species in focus here.

To study how different species – in this case humans and mainly either bear or wolf – interact, requires a basic understanding of all sides. It is necessary to recognize that the encounter is reciprocal and that there are at least two independent agents involved. Whether the carnivores are actually present is of course important. In another article I have shown how the narratives about wolves changed as the animal disappeared from the wild.[1] This was also true of the bear and other large predators. Generally, the stories grew more elaborated and extended as the animals became rare in the wild. Discussing the reciprocal encounters between human and non-human animal demands considering the agency of the counterparts. The agency of non-human animals has been addressed broadly within the last decade in humanistic disciplines. There are basically two strands to the discussions. On the one hand the debate has been inspired by Bruno Latour and actor-network theory.[2] The idea here is to recognize how objects can influence humans and in this way function as actors with meaningful participations in history. On the other hand, the focus has been on non-human animals as agents, as intentional beings co-evolving with humans in history. The latter strand is mainly inspired by Donna Haraway's work.[3] This chapter is aligned with the Haraway's way of thinking about the co-evolution of human and non-human animal.[4] The agency of the large carnivores acting in the narratives examined here is considered to be an important factor in history. I believe animals to be intentional beings. However, in human accounts this independence is often disguised and the non-human animals are objectified. The dominance of the human over the non-human is often the focus of the account but this is merely one way of seeing things. The story below can be read as a report on human dominance but it can also be interpreted as a tale of how perceived agency in non-human beings prompts humans into elaborate and intricate behavioral performances. It can also be read as a story about how animal agency is constructed in historical texts.

This perspective puts the historian in a difficult position especially in regard to source material. If non-human animals are to be considered powerful agents and their views demand to be taken into account, the question remains: where do we find source material telling a non-human history? Is the potential agency found in historical texts a reflection of animal intentionality or a human construct? The point of departure of this chapter is the consciousness of our lack of information on this matter. It is important to be aware of the multitude of stories, which are not obvious in ordinary historical source material. The non-human animals acting in the stories conveyed below are mainly trying to evade hunters. They are frequently forced to participate in performances directed by humans.

It is only when human and carnivore meet that the agency of the animal becomes visible. The non-human agency between these events is much harder to approach. It is, however, important to recognize that what brought the animal to the point when it encountered the human is quite a different story, not necessarily known to us, and not possible to address here. It is, however, equally important to recognize animal agency when it appears in the source material. The idea of animal agency can in this way be a method to be aware of the presence of these stories, and at least to partially hear them being told. The belief in that there are stories hidden in history from tellers who are not able – or even willing – to convey them, enables us to accept a wider perspective and to take a somewhat humbler approach to the non-human animals appearing in history. Accepting the agency of non-human animals means understanding and considering that the animals have life-worlds of their own and that their lives are not only influenced by us but also have a deep impact on ours.[5]

The idea of this chapter is thus to take the point of departure in the reactions awakened in humans by the large carnivores and how they are expressed in accounts on encountering them. That way the focus will be on how the carnivores influence us and not the other way around. The responses addressed here are exclusively human but behind them we can trace non-human animals' behavior. The movements and actions of the carnivores are not further discussed here although they certainly also changed during the period. The time frame of the article – the nineteenth century – corresponds to the period when the populations of large carnivores were subsequently decimated in Sweden. At the end of the period the animals acting in these narratives increasingly became fictitious. In this way the agency of the animals sometimes reflects actual actions by animals, and sometimes is constructed by humans. The source material consists of Swedish literature on hunting and natural history from the nineteenth century as well as notes and articles found in the royal Swedish library's digitalized newspapers from the same period.

Folklore and Zoology

The bear was the most admired of the four large carnivores existing in Sweden at the time. Though it was the source of much hatred and fear it maintained an aura of respect. It was, since times immemorial, considered to be a humanlike creature. Stories about bears were common in folklore and were often considered trustworthy.[6] In Nordic folklore there was a widespread association between large carnivores and sexuality. It was thought that pregnant women were particularly at risk of being assaulted by bear or wolf.[7]

Leading zoologists such as Sven Nilsson (1787–1883) referred to folklore in order to characterize the animals.[8] Nilsson was noted for the breadth of his multidisciplinary scholarship, and not the least for his great inventory of the Scandinavian fauna first published in 1820 and thereafter revised and published again in 1847.

The natural history of the bear was inherited knowledge from the eighteenth-century writers. In their work the bears were often considered to be of different kinds, ordered by their perceived diet. There was the larger and more dangerous killer bear and the smaller, more peaceful ant bear. The different sorts were recurrent in natural history, sometimes completed with a third kind: the grass bear. The bear was considered to be herbivore or semi-carnivore. It will sometimes bother livestock, Nilsson stated, but most often it is the livestock which provokes an attack from the bear, by showing aggression towards it.[9] The moral superiority of the bear, compared to other carnivores, seemed mainly to be linked to its diet. As a semi-carnivore the bear was thought of as more capable of controlling itself than other predators (Figure 12.1).

The encounter between a bear and a human influences both, stated Nilsson. The dominance of humans over animals made the bear confront the human in a

Figure 12.1 Herman Falk was aware of, and acknowledged, the agency of the bear. This bear, on the front page of Falk's hunting instructions, looks wearily into the eyes of the hunter.

Source: Falk, *Underrättelser om Björn Skall.*

different mode than other animals, according to the zoologist. The human would be mauled while the bear would strike other animals with its paws. A great force of power, argues Nilsson, lies in the gaze of the human.[10] Killing a bear was considered a manly deed. It was thought, reported one author, that courage could be received by drinking the bile of the killed bear.[11]

During the nineteenth century natural history changed in Sweden as elsewhere. The system of Linnaeus, which at the time was a counterpart of the romantic biology with advocates such as Elias Fries, was towards the second half of the century replaced by a more dynamic evolutionary way of thought, culminating in the teachings of Charles Darwin. Towards the end of the nineteenth century the Darwinian theory was subsequently accepted. Darwin's theory was in Sweden first acknowledged in the institutions, such as the Museum of Natural History or the Royal Academy of Sciences, not dominated by older natural history.

The animals in zoology thus increasingly became placed in an environment, in geography and in relationship with their surroundings.[12] There was a change – probably the consequence of the leading zoologists' interests in evolutionary thinking and geography – towards describing the animals as parts of a larger context, a landscape or a culture. A broad interest in evolutionary thinking in the widest sense emerged, where one asked what people or what fauna had been present in the country earlier, and how the present time situation had evolved. The animals, including large carnivores, thereby became part of an evolutionary history where they also had a place in the national history.

History and evolution thus converged in these narratives and anthropology and zoology fertilized each other to a greater extent than earlier. In this way descriptions of the animal's geographical prevalence, not merely their habits, use and appearance became common in zoology. During the nineteenth century animals thus became linked to a national self-image and thereby also could be seen as belonging to *the Swedish*. In a popular book on zoology, published in 1903, the animals were presented and organized completely in accordance with their habitat.[13]

The idea of the carnivores as part of the Swedish fauna was new to nineteenth century Sweden. Earlier large carnivores were pursued based on the argument that they were not Swedish. Frequently – in folklore as well as in early popular zoology – threatening carnivores were linked to others, mainly the Sami people or Finns. At the beginning of the nineteenth century there was a conception of increasing wolf predation, which was commonly explained as being Russians turning prisoners of war into wolves.[14] A campaign was launched during the eighteenth century to drive large carnivores over the nation's borders and instructions on how to kill carnivores were conveyed in every church in the country by order of the parliament.[15] In this way large carnivores were externalized and described as foreign.

Though the system and form of natural history clearly had changed during the nineteenth century, the contents and narratives were very much the same as before. The basic knowledge of large carnivores, gained from folklore and older natural history, was maintained. The negative connotations and associations attached to the animals remained. Sometimes they lingered in a dialogical sense, as a notion to contest, and sometimes they were explicit.

In this context the carnivores were commonly ascribed agency. Their actions were depicted as outcomes of their perceived traits. The bear would act noble and the wolf would display greed, etc. Such implicit, dialogical ideas were quite common in natural history about large carnivores. The dialogue was usually held with recognized, well known but not written, knowledge. In the different genres discussed in this chapter, all concerned with natural history and wildlife, knowledge was transferred quite freely from folklore, to fact, to fiction, and back again.

Hunting Stories

Knowledge about the bear was in this way impregnated with ideas about its powerfulness. This idea probably was reinforced by the agency of actual bears being under attack. With remarkable frequency the hunting stories concerning bears were about female bears and their cubs being attacked in their dens. This was certainly a situation where the desperation of the bears could increase the view of the animal as powerful and dangerous.

To bring down a bear was considered admirable, whether the animal was juvenile or full grown. Hunting stories about bears can be considered a subgenre with the huntsman Llewellyn Lloyd (1792–1876) as the Nestor. Lloyd was an Englishman who came to Sweden and Norway in the 1820s to fish and hunt. His enchantment with the Nordic countries, his independence of means, and possibly his need to evade the English juridical system after accidentally having shot a forest officer while poaching, made him stay and publish hunting stories and accounts of his encounters with the Swedish peasantry.[16] Lloyd described bear hunting as a colourful and dangerous operation.

The admiration of the bear made the hunt appear as a royal performance with the bear as a revered participant. During the nineteenth century battues for large carnivores were quite frequently narrated as instructive stories about the organization of the hunt. Lloyd received his training as a huntsman from the famous bear hunter and royal forest officer Herman Falk (1785–1865).[17] As so many of the professional hunters of the time, Falk began as a military man. After resigning in 1812 he became famous for his skill at organizing bear battues. He was a good friend of the zoologist Sven Nilsson and provided the professor with knowledge about bears and their nature and way of living.[18] Falk was admired for his expertise in organizing battues with large amounts of people. In his writings on the matter, a carefully choreographed and ritualized dance – involving both bears and people – emerges. Falk showed awareness about how wild animals react. His battues displayed a professional performance of men as well as bears. In this dance, in his account illustrated with a map, the "beaters," driving the animal, were separated from the "holders" and all were marked with a specific number corresponding to a specific place in the line.[19] See Figure 12.2.

The drivers must start simultaneously walking very slowly and in complete silence, Falk instructed.[20] Irregularities or sharp edges of the line of beaters brought great danger, because it made it easy for the bear to break through.[21]

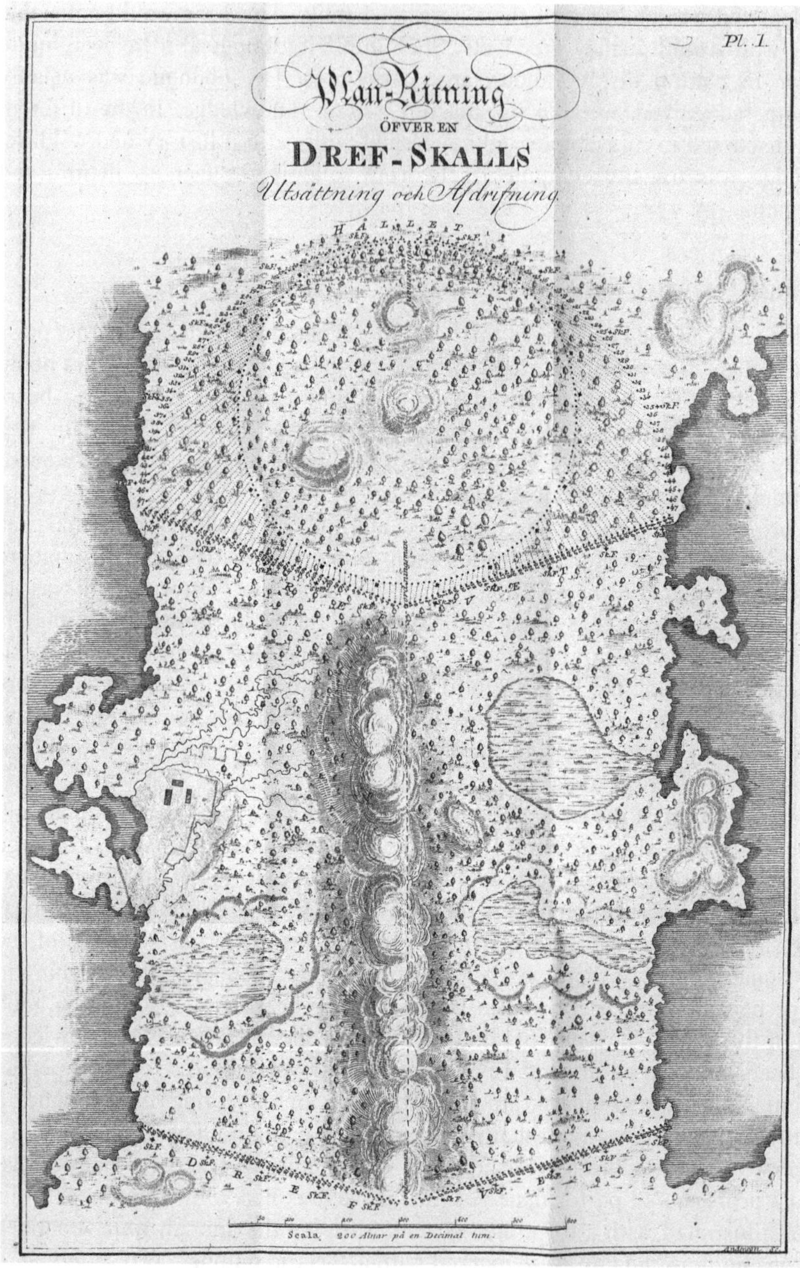

Figure 12.2 A detailed map accompanied Falk's hunting instructions. Every person is marked with a number corresponding to a specific place in the line.

Source: Falk, *Underrättelser om Björn Skall*.

The silence of the search party was necessary in order to keep the bear calm. Should the drivers start shouting or making other noise the bear might easily panic and run through the line. Noise should only be made, Falk stated, if the bear tried to escape. But the moment he turns back, Falk instructed, complete silence must be observed.[22] The aim of the hunting operation was to slowly and calmly drive the bear towards the gunmen.

The methods of hunting used during the nineteenth century were in this way often expressed in terms of order and organization. The agency of the bear was acknowledged in the way that the excessive order and ritual of the hunt was viewed as necessary in order to manage the animals. The will and the intentions of the animal were seldom addressed but the whole operation was designed to overcome them.

Falk expressed the idea that bears had to be treated with skill not to become dangerous. He knew, and acknowledged the fact, that neither the bears nor most of the people were voluntary participants of the battue.[23] The aim of his precise instructions on how to perform a successful battue was to force both animals and humans to move in a predictable way. The whole operation was based on the acceptance of the animals as intentional beings, trying their best to avoid the hunters.

The ceremony and ritualistic performances of the royal hunting practices have been widely discussed in research.[24] The main interest for the royalty was the actual killing of the animal and a sign of the skill of the royal forester and his performance was his ability to keep the wild bears in place for the king to kill them. A good hunt was so well arranged that both animals and manpower performed in accordance with the will of the king. The display of power remained an important aspect. The kill was the climax of the hunt. The king, or someone who was permitted by him, performed it.[25]

Threatening Agents

Choreographing the hunt thus meant being knowledgeable about the nature of the animals hunted. Though the order of the hunt made the animals move according to plan there was, during the nineteenth century, a growing awareness of the agency of the animals. The chief forester J. D. Arrhenius, in 1833, stressed the importance of knowledge of the animal to be able to hunt it. The hunter must know where the animal wants to go, what it likes and how it finds what it wants, to be able to hunt it successfully.[26]

The agency of the wolf was also enforced by the way it was described in newspapers. The wolf most often "showed itself" or "visited."[27] Occasionally reports about wolves attacking humans appeared in the newspaper. One paper in 1861 told about a "pugnacious" wolf biting both a farmer and his neighbor before being shot. The wolf, according to the newspaper, approached the farmer unprovoked and bit him.[28] Wolves were frequently described as brave and bold enough to come close to people.[29] Wolves appearing in newspapers were more often than not described as "large" or even "exceptionally large."[30]

The animals were in this way most often described as daring, being brave or attacking. The animal's agency was expressed in terms of it being unruly and disorderly. Some wolves were portrayed as experienced and therefore "difficult and unstable" to handle thus being very challenging to catch and keep within the nets.[31] Some animals were "spoiled" and would repeat a successful strategy to evade hunters.[32]

This picture was enforced in natural history where the wolves were described as an uncontrollable threat. In the ideas inherited from the eighteenth century wolves seemed to flood the country. They appeared to invade from Norway, Finland and Russia.[33] In works by Sven Nilsson this view was further developed and he described the wolves as immigrants from Finland and Russia.[34]

During the first half of the nineteenth century the object of the great hunts more and more became focused on the eradication of large carnivores. The aim was to drive them from the country. The large hunting operations were increasingly described as a duty, and less as a pleasure for royalty. The great royal hunting performances required enormous amounts of manpower and resources. Therefore they were also widely criticized.[35] When the large hunting operations were no longer considered fashionable, efforts were made by the authorities to encourage the general public to hunt large carnivores with greater efficiency than before. Statistics and information about animals killed were published in order to inspire extermination of predators.[36]

A New Hunter Emerges

The official management of large carnivores, developed during the nineteenth century, was thus mainly aimed at eradicating them or driving them away.[37] The nineteenth-century nationalism linked the operation to the concept of the Swedish. Therefore, the hunting of large carnivores came to be seen as the main task for professional hunters. It was their duty to perform wildlife management, which was the equivalent of eradicating pest animals. The hunter was described as a protector of national interests.

The cost of predation of large carnivores on livestock was calculated and frequently published in newspapers. The relationship between human and carnivores was increasingly described in economic terms.[38] The arguments for killing large carnivores were thus expressed as a concern for the general public and the whole nation.[39] Influential in this aspect was, again, the professor from Lund, Sven Nilsson.

Nilsson's fauna is rich with observations made by hunters and stories about great hunting adventures. Nilsson's double interests, as a scientist and as a hunter, indicate a change in the nineteenth-century relationship with wild animals. Hunting, which earlier had been performed as a military exercise and a display of power, now contained elements of gazing and observing. Animals were increasingly considered a part of nature, inhabitants of the geography. The hunter was expected to have more knowledge about them than was needed to catch them. Hunting was during the nineteenth century in Sweden also established as a practice and a pastime for other groups than the nobility. The nobility and the royalty continued to show interest in hunting. Hunting for pleasure had always, and particularly since the large debate about hunting laws in 1734, been associated with status.[40] Therefore an upcoming bourgeoisie took to hunting as a means to establish itself. The founding of a society for those interested in hunting and natural history and the publishing of a new journal on the matter of hunting should be viewed in this perspective.

In the Swedish hunting journal there was an obvious strife to portray hunting as an admirable, historically significant and heroic task. The journal wanted to establish hunting as an occupation with specific connotations. It was considered important to define hunting as a concept and to identify the hunter. In the journal it was stated that the roots of hunting were to be found in times immemorial when humans needed to expand their territory. Hunting was originally thus a warfare fought to acquire new territory, the journal explained. The purpose of hunting was to conquer the inaccessible wilderness. This land was at the time occupied by wild animals which humans were forced to declare war against.[41] The colonial usage of language could hardly be more obvious.

In the Swedish parliament hunting was quite frequently discussed and the aim was most often to encourage a more effective hunt for large carnivores. The presentation of the relationship with large carnivores as an economic one, where the presence of large carnivores was seen as a cost, fit well into this development. The object of the hunt was to protect the nation, and to protect it from foreign or costly elements.

Understanding Large Carnivores in Literature

Both hunters and naturalists thus increasingly, during the century, emphasized the importance of knowledge when encountering wild animals. To know the carnivore one was required to adopt its perspective, feel what it feels and smell what it smells. This inside perspective was a part of the important and necessary knowledge of the hunter. It was recognition of the agency of the animals. It opened up a new viewpoint, a possibility to see the world from the animal's angle. Knowledge was necessary to predict the animal's actions. This knowledge, however, also involved an empathic view of the carnivores. It meant seeing the animal from within, to understand its attempts to live its life, protect its young and go on with its business. This perspective subsequently found its way into the late nineteenth-century literature with writers such as Selma Lagerlöf.

Lagerlöf was familiar with the works of Sven Nilsson. The link between animals and geography – essential in the development of nineteenth century natural history – was central in Lagerlöf's writings.[42] Her instructive geography schoolbook, *The Wonderful Adventures of Nils* (published 1906–1907) became the literary symbol of the link between nationalism, animals and geography. The boy, Nils, literarily is forced to adopt the animal's view of Sweden when he is turned into a manikin and as such travels with the migrating wild geese through Sweden.[43] In this way even large carnivores came to be seen as belonging to the Swedish nation.

Adopting the perspective of the animal was also probably important for the growing ecological awareness. Towards the twentieth century there was a growing idea about the necessity of balance in nature, a concept of a system in which all species were necessary. This idea was first put forward in Sweden by the archaeologist Per Arvid Säve in an article in the Swedish hunting journal.[44] In the whole of the western world there was a growing interest in protecting threatened species and values found in nature. By the end of the nineteenth century there was for instance, as a consequence of this in Sweden, a growing opinion wanting to protect the bear.[45]

A critical aspect in relation to nature emerged in literature towards the end of the nineteenth century. In Selma Lagerlöf's debut *The Story of Gösta Berling* (1891), the tale about the bear of Gurlitta klitt becomes a prism of the changing attitudes towards animals in nineteenth-century Sweden. In describing the hunt for the bear Lagerlöf suddenly shifts the perspective to the bear and we experience its irritation when the hunters wake it in its den. When the bear charges through the entrance of the den, it sees the hunter, Gösta Berling himself, who is unable to fire his weapon. This is because he too experiences the perspective of the bear. The agency of the bear becomes forcefully obvious to Gösta Berling, who now sees the animal as what it is, a poor, hunted creature. He is reluctant to bereave the animal of the last thing it possesses – its life – after the humans have taken everything else away from it. Gösta Berling shows empathy with the bear. This was new to the hunting literature of nineteenth-century Sweden.

The critique of civilization of the time made it possible to both describe the hunt and at the same time criticize it. Such is the story by the painter, journalist and writer of short stories, Pelle Molin. Molin's short stories were published after his early death in 1896. The story about the settler Salmon, who encounters a bear when running through the forest at night in pursuit of a midwife to help his wife who is in labor, contains such a double-edged critique. The story tells how the man and the bear run around a tree, the man wondering about the source of the rage of the bear and at the same time worrying about his wife left alone at home with three kids and the fourth on the way. Finally, Salmon manages to brutally kill the bear with his knife, only to find it was merely protecting its cub, hiding in the branches of the circled tree. The thought dawns on Salmon that the encounter between himself and the bear has bereaved them both of their next of kin.[46] Salmon is reduced to the *nature* he is trying to battle. He and the bear share a common problem. He is captured in a structure neither he nor the bear are able to shake off.

Molin's description of the bear encounter was, just like the earlier hunting stories, expressed in terms of a dance. Towards the end of the nineteenth century the large battues disappeared but were replaced by a different kind of dance. The title of the short story "A ring dance while mother waits" refers to both the killing of the bear and the structure entrapping both bear and human. The dance was now less tightly choreographed, involved merely two counterparts and was described as an event taking place between two independent agents acting voluntarily. The hunt was, in the beginning of the twentieth century described as a game or a sport where the hunter invites the animal to take part.[47] In Molin's story, however, both the human and bear are ensnared.

The Affection

Though killing large carnivores went without saying, the actual act of killing seems to be surrounded by anxiety. The actual methods were conveyed in instructive hunting literature but when the event was narrated the kill was expressed with euphemisms. The animal was said to "pay with its life" or "bite the grass" and the act of the hunt was described as "getting rid of unwanted guests" or "shortening the days of the animal" and so forth.[48]

Many stories told about encounters with large carnivores were also focused on love and affection. Wild animals, including carnivores, could be kept for amusement. Animals would be caught and held for longer or shorter periods of time. Domesticated bears were not unusual. The huntsman Llewellyn Lloyd was often portrayed walking his pet bear and the hunter and author Gustaf Schröder (1824–1912) in his writings expressed delight with his.[49]

The most famous, or notorious, of captured carnivores was the so-called Gysinge wolf, probably responsible for the only documented wolf killing of children in modern times in Sweden. A triplet of wolf cubs was captured and raised by the Bedoir family in Gysinge. Two of the cubs were given away and the Bedoir family kept the third. The following year a series of wolf attacks occurred in the neighborhood, leading to the conclusion that the semi-domesticated wolves were responsible. Whether this was the case or not will never be known, however Carin Bedoir, the lady of the house, claimed their wolf was shot at the age of one year.[50] What happened to the other two cubs is unknown.[51] The wolf killings of 1820–1821 were also mentioned in the newspaper.[52]

The carnivores were thus encountered with mixed emotions. Even the famous bear slayer Llewellyn Lloyd expressed his concern for the bears being exterminated by the relentless Swedish hunting practices.[53] At the end of the century there was a growing fascination with relationships between humans and animals. Stories about encounters between humans and wild animals became abundant in press as well as in children's literature.[54]

The nineteenth-century Swedes seemed very attentive to the blurring of the nature–culture divide. The changing narratives about large carnivores thus reflect a changing attitude towards nature in Sweden at the time. Earlier, nature was considered to be an enemy to be defeated. Wild animals, and especially large carnivores, were to be eradicated as enemies of the nation. Eventually nature rather became conceived as equivalent to the *natural* and therefore a place to learn from and to understand.

Conclusion

The frameworks in which the large carnivore encounters were narrated thus changed during the nineteenth century. The process was complicated and double-edged and different tendencies can be noted. On the one hand there was an increasing closeness between human and non-human. The choreography of the large hunting operations developed into an individual dance where the animal was described as a voluntary participant. The military operation continuously, during the century, gained elements of observing and gazing. The hunter became a guest, and a seeker of knowledge, in nature. The previous gap between nature and culture began to close, allowing nature to be perceived as *natural*. These aspects of the stories of encountering large carnivores all pointed towards a closer relation between human and carnivore. It became possible, and desirable, to adopt the animal's perspective. To be able to hunt the animal one must become familiar with it. This closeness was also conveyed in literature.

On the other hand, the hunt was also articulated as an alienation of carnivores from humans. As the century progressed they were to a greater extent described as threatening. They were in this process ascribed more agency but it was narrated in terms of threat. During the century the relationship between humans and carnivores became increasingly described in economic terms. The way of thinking fit well into the discourse differing useful animals from harmful ones, inherited from the eighteenth century. It also fit into the eighteenth century wartime rhetoric where the battle against large carnivores was described and performed in terms of military operations.[55] Harmful animals – mainly large carnivores – were seen as being not merely morally detestable but also enemies of a growing nation. Towards the second half of the century, however, the idea emerged of the carnivores as being part of the Swedish nation, and thus worthy of protection. The carnivores were portrayed in context, in nature and as a part of the Swedish nation. This way of thinking about the animals structured fact as well as fiction.

The ascribing of agency to the carnivores worked in both ways. As the animals were no longer controlled in the organized large hunts they could to a greater extent be described as threatening and dangerous. They were frequently depicted as being offensive, approaching or attacking, despite the fact that the number of actual animals in the wild had plummeted during the century. This indicates that the ascribed agency mainly was a construct, aimed at the management of large carnivores.

The agency of the non-human animals was in this way articulated in an indirect fashion. It can be read into the human reactions towards the carnivores. The humans acted as if the non-humans were independent agents. Nevertheless, their perceived agency was not directly addressed in the beginning of the century. The large hunting operations were designed to manage independent animals with a will of their own but the agency of the non-human participants was either disguised or ignored.

Towards the second half of the nineteenth century the populations of large carnivores in Sweden were pushed north and decimated. Parallel to this process the agency of the carnivores became articulated to a greater extent. As the real animals disappeared, fictionalized agents appeared in newspapers and literature. By the end of the nineteenth century the idea of the value of wild animals in nature was well established. Due to the work done by the hunting organization as well as the development within zoology and, not the least, the empathic adoption of the non-human perspective in literature, it became possible to argue for the protection of large carnivores. This, however, in practice only concerned the bear, and subsequently the lynx. The feared wolf and the detested wolverine were not considered worthy of protection until the late twentieth century.

Notes

1 Dirke, "Where Is the Big Bad Wolf?"
2 Latour, *Reassembling the Social.*
3 Haraway, *When Species Meet.*
4 Haraway, *Companion Species Manifesto*, 12.
5 The term "life-world" is taken from the Estonian biologist and philosopher Jakob von Uexküll who used it as a way of penetrating and understanding the lives of animals. Uexküll, "A Stroll through the Worlds of Animals and Men," 5–80.
6 Bernström, *Bernströms bestiarium*, 50.
7 Tillhagen, *Jaktskrock*, 135; Ekman, *Herrarna i skogen*, 194–8.
8 Nilsson, *Skandinavisk Fauna*, 197f.
9 Nilsson, *Skandinavisk Fauna*, 203. This was an idea Nilsson received from Falk. See Falk, *Underrättelser om BjörnSkall*, 28. The idea of the bear as herbivore is conveyed also by Wetterhoff. See Wetterhoff, *Från skog och sjö*, 10.
10 Nilsson, *Skandinavisk Fauna*, 199. The famous bear hunter Herman Falk also conveyed this idea however without the link to the human gaze.
11 Wetterhoff, *Från skog och sjö*, 9.
12 Wirén, *Zoologiens grunddrag.*
13 Jägerskiöld, Lönnberg and Adlerz, *Sveriges djurvärld*, 198.
14 Tillhagen, *Jaktskrock*, 160.
15 von Liewen, *Wetenskap at döda Wargar.*
16 Björklöf, *Björnen*, 167. According to Björklöf he arrived in Sweden as early as in 1819.
17 Petre, "Lloyd, Llewellyn," 65.
18 Amenius, "Falk, Herman Adolph," 484.
19 Falk, *Underrättelser om Björn Skall*, 4.
20 Ibid., 9f.
21 Ibid., 11.
22 Ibid., 12; *Tidskrift för Jägare och Naturforskare*, no. 4–5, 1833, 496.
23 Falk, *Underrättelser om Björn Skall*, 4, 12.
24 See for example, Nyrén, "Rätt till jakt," 24–5, 212–3. Nyrén views the hunting practices of the elites in Sweden as attempts to match international performances but also stresses the Swedish peasantry's opposition by ignoring hunting laws.
25 *Tidskrift för Jägare och Naturforskare*, no. 1, 1833, 389.
26 *Tidskrift för Jägare och Naturforskare*, no. 4–5, 1833, 487.
27 See for example, *Kalmar*, January 19, 1867, 2; *Kalmar*, December 1, 1869, 2; *Blekingsposten*, June 13, 1854, 1. In *Dalpilen* the wolves were described as intrusive. See *Dalpilen*, January 13, 1866, 3. In the same newspaper, in an annual report on the wild game in Sweden, the wolf was described as "especially bold" despite the mild winter. *Dalpilen*, April 14, 1866, 2. See also *Jönköpingsbladet*, March 14, 1865, 2. The same story also appeared in *Dalpilen*, March 18, 1865, 2.
28 *Tidning för Wenersborgs stad och län*, February 7, 1861, 1. A series of wolf attacks on children in Finland were reported in 1876–1881. These reports are further discussed in Dirke, "Where is the Big Bad Wolf?"
29 *Jönköpingsbladet*, March 18, 1851, 2.
30 See, for example, *Jönköpingsbladet*, December 14, 1850, 1; *Jönköpingsbladet*, January 17, 1861, 2; *Dalpilen*, January 13, 1866; *Kalmar*, January 23, 1871, 2;

Faluposten, January 25, 1871; *Dalpilen*, March 13, 1875, 3; *Faluposten*, February 26, 1876, 2; *Faluposten*, March 16, 1878, 2; *Kalmar*, January 8, 1890, 4; *Kalmar*, May 10, 1890, 2.
31 *Tidskrift för Jägare och Naturforskare*, no. 4–5, 1833, 486. The Swedish expression is "kinkige och ostadige."
32 Ibid., 495.
33 Berch, "Vestmanlands björn-och vargfänge," 38.
34 Nilsson, *Skandinavisk fauna*, 223.
35 Hagelin, *Samling af handlingar*.
36 Bergström, Dirke and Danell, "The Wolf War in Sweden."
37 Bergström, Dirke and Danell, "The Wolf War in Sweden."
38 See, for example, *Göteborgs Weckoblad*, May 12, 1881, 1; *Dalpilen*, May16, 1884, 5; *Norra Skåne*, March 31, 1888, 1; *Norra Skåne*, October 24, 1895, 2. In *Norra Skåne* on October 24 it was also noted that the value of the killed livestock is higher than the paid bounties.
39 *Faluposten*, September 11, 1875, 3; *Göteborgs Weckoblad*, May 12, 1881, 1.
40 The debate has recently been thoroughly investigated by the historian Ulf Nyrén in his dissertation. See Nyrén, "Rätt till jakt." For an older work, see Östergren, "Tvisten om ägande- och nyttjanderätten."
41 *Tidskrift för Jägare och Naturforskare*, no. 1, 1833, 385.
42 Edström, *Selma Lagerlöfs litterära profil*, 91–132.
43 For a discussion on the nationalism in Lagerlöf's book, see Sundmark, "Of Nils and Nation."
44 Säve, "Sista paret ut!" 70–86.
45 Jägersköld, Lönnberg and Adlerz, *Sveriges djurvärld*, 22.
46 Molin, "En ringdans medan mor väntar," 82.
47 Möllersvärd, *Med stövare i skogen*, 32; Eklundh, *På jakt*, 33, 190.
48 Examples of such euphemisms are found in *Göteborgs Weckoblad*, May 12, 1881, 1; *Dalpilen*, October 24, 1899, 3; *Kalmar*, March 5, 1877, 3; *Jönköpingsbladet*, December 19, 1863, 3; *Post- och Inrikes tidningar*, July 7, 1856, 2; *Post- och Inrikes tidningar*, March 4, 1844, 2.
49 Björklöf, *Björnen i markerna och kulturen*, 166–7, 219–23.
50 *Tidskrift för Jägare och Naturforskare*, no. 8, 1832, 243ff.
51 Ekman, *Vargen. Den jagade jägaren*, 202.
52 *Fahlu Weckoblad*, January 20, 1821, 2.
53 Lloyd, *Jagt-nöjen i Sverige och Norrige*, 21, 32.
54 Dirke, *De värnlösas vänner*, 142–55.
55 Bergström, Dirke and Danell, "The Wolf War in Sweden."

Bibliography

Newspapers

Blekingsposten, 1854.
Dalpilen, 1865, 1866, 1884, 1899.
Faluposten, 1871, 1875, 1878.
Fahlu Weckoblad, 1821.
Göteborgs Weckoblad, 1881.
Jönköpingsbladet, 1850, 1851, 1861, 1863, 1864, 1865.
Kalmar, 1867, 1869, 1877, 1890.
Norra Skåne, 1888, 1895.
Post- och Inrikes tidningar, 1844, 1856.
Tidning för Wenersborgs stad och län, 1861.
Tidskrift för Jägare och Naturforskare utgifven af Jägare-förbundet, 1832, 1833.

Literature

Amenius, Ragnar. "Falk, Herman Adolph." In *Svenska män och kvinnor: biografisk uppslagsbok*, edited by Nils Bohman and Torsten Dahl, 484–485. Stockholm: Bonnier, 1942–1955.

Berch, Anders. "Vestmanlands bjirn-och varg.fänge, ... under ... Anders Berchs biträde, förestält af Anders Hillerström ... uti then större Carolinske lärosalen den [23] Maji år 1750." PhD diss., Uppsala universitet, 1750.

Bergström, Roger, Karin Dirke, and Kjell Danell. "The Wolf War in Sweden during the Eighteenth Century – Strategies, Measures and Leaders." In *A Fairytale in Question. Historical Interactions of Humans and Wolves*, edited by Patrick Masius and Jana Sprenger, 57–78. Isle of Harris, UK: White Horse Press, 2015.

Bernström, John. *Bernströms bestiarium: en djurens nordiska kulturhistoria*, edited by Henrik Otterberg, 48–53. Stockholm: Atlantis, 2008.

Björklöf, Sune. *Björnen i markerna och kulturen*. Stockholm: Gidlunds förlag, 2010.

Dirke, Karin. "Where Is the Big Bad Wolf? Notes and Narratives on Wolves in Swedish Newspapers during the 18th and 19th Centuries." In *A Fairytale in Question. Historical Interactions of Humans and Wolves*, edited by Patrick Masius and Jana Sprenger, 101–118. Isle of Harris, UK: White Horse Press, 2015.

Dirke, Karin. *De värnlösas vänner. Den Svenska djurskyddsrörelsen 1875–1920*. Stockholm: Almqvist & Wiksell International, 2000.

Edström, Vivi. *Selma Lagerlöfs litterära profil*. Stockholm: Rabén & Sjögren, 1986.

Eklundh, Albert. *På jakt: jaktskildringar*. Stockholm: Norstedt, 1926.

Ekman, Henrik. *Vargen. Den jagade jägaren*. Stockholm: Norstedt, 2010.

Ekman, Kerstin. *Herrarna i skogen*. Stockholm: Bonnier, 2007.

Falk, Herman Adolph. *Underrättelser om Björn Skall*. Stockholm: H. A. Nordström, 1828.

Hagelin, Magnus. *Samling af handlingar om nödige anstalter emot rofdjurens förökelse. Den 2 julii 1798. Til konungens befallningshafwande om ludringar och andra medel emot odjuren*. Linköping: D. G. Björn, 1800.

Haraway, Donna. *When Species Meet*. Minneapolis, MN: University of Minnesota Press, 2008.

Haraway, Donna. *Companion Species Manifesto: Dogs, People and Significant Otherness*. Chicago: Prickly Paradigm, 2003.

Jägerskiöld, L. A., Einar Lönnberg, and Gottfrid Adlerz. *Sveriges djurvärld*. Stockholm: Ljus, (1903) 1911.

Latour, Bruno. *Reassembling the Social. An Introduction to Actor-Network-Theory*. Oxford: Oxford University Press, 2005.

Liewen, Berent Wilhelm von. *Wetenskap at döda Wargar. Kan ock dödas andra Rofdjur och Rof-foglar*. Stockholm: tryckt uti kongl. tryckeriet, 1753.

Lloyd, Llewellyn, *Jagt-nöjen i Sverige och Norrige: berättelser om en resa i dessa riken åren 1827–28*. Stockholm: Hjertas, Wiborgs & Comp, 1830.

Molin, Pelle. "En ringdans medan mor väntar." In *Ådalens poesi*, 82. Stockholm, Wahlström & Widstrand, 1897.

Möllersvärd, John. *Med stövare i skogen. Jaktskildringar*. Stockholm: Åhlén & Åkerlund, 1920.

Nilsson, Sven. *Skandinavisk fauna. En handbok för Jägare och Zoologer*. Lund: Gleerup, (1820) 1847.

Nyrén, Ulf. "Rätt till jakt: en studie av den svenska jakträtten ca 1600–1789." PhD diss., Göteborgs universitet, 2012.

Östergren, P. A. "Tvisten om ägande- och nyttjanderätten till skattejord vid pröfningen af Lagkommissionens förslag till Sveriges Rikes Lag hos 1731 och 1734 års ständer." PhD diss., Lunds universitet, 1896.

Petre, Allan. "Lloyd, Llewellyn." In *Svenska män och kvinnor: biografisk uppslagsbok*, edited by Nils Bohman and Torsten Dahl, 65. Stockholm: Bonnier, 1942–1955.

Sundmark, Björn. "Of Nils and Nation: Selma Lagerlöf's *The Wonderful Adventures of Nils*." *International Research in Children's Literature* 1 (2008): 168–186.
Säve, Per Arvid. "Sista paret ut! (tankar om behofvet af en lag mellan menniskan och djuren)." *Svenska Jägareförbundets Nya Tidskrift* 2 (1877): 70–86.
Tillhagen, Carl-Herman. *Jaktskrock*. Stockholm: LT, 1985.
Uexküll, Jakob von. "A Stroll through the Worlds of Animals and Men: A Picture Book of Invisible Worlds." In *Instinctive Behavior. The Development of a Modern Concept*, edited by Claire H. Schiller, 5–81. New York: International Universities Press, 1934.
Wetterhoff, Onni. *Från skog och sjö. Jagt- och fångstfärder*. Stockholm: Hæggström, 1883.
Wirén, Axel. *Zoologiens grunddrag*. Stockholm: Bonnier, 1899.

13 The Wise Salmon That Returned Home

Outti Autti

Introduction: Salmon in Two Northern Rivers

In this chapter, I will analyze interview data that I collected among salmon fishermen living along two rivers in Northern Finland, the Iijoki and the Kemijoki. The fishing stories of the interviewees go back to the 1920s and end with the damming of both rivers. I analyze the variety of features associated with salmon, and investigate how or why the interviewees justified these associations. By observing how the fish were represented and described in the interviews, this chapter seeks to understand the essence of salmon agency and the human–salmon relationship.

Local cultures in Northern Finland, like elsewhere in the Circumpolar North, have a long history of relying on migratory fish for their sustenance. Atlantic salmon (*Salmo salar*) have played a major role in the history and development of societies and cultures in northern Europe. Atlantic salmon typically hatch in the gravel of a freshwater riverbed, then spend several years in the river. They then migrate to the ocean where they stay several (1–5) years, after which they return to the river habitat to spawn.

The annual salmon migration to freshwater rivers has strongly shaped the northern riverside cultures. Fishing migratory fish was an important source of livelihood and one of the main things that historically drew permanent settlements to the northern riversides. Besides their nutritional value, there were many other important characteristics of migratory fish, including aesthetic, symbolic, social, and economic values. Migratory fish and their annual rhythm belonged to the rivers' landscapes, defining the annual rhythm of riverside villages that adapted to the salmon migration.[1] Specific professions and skills were developed and buildings were constructed to serve salmon fishing. Salmon shaped people's ways of life and their thinking.[2]

The Kemijoki was one of the most important salmon rivers in Europe,[3] while the Iijoki was one of most important salmon rivers in Finland.[4] Salmon started the climb to their breeding areas annually in June, but fishing continued until autumn with the migratory whitefish. The fish were prepared for food, sold, or stored for later use. Salting or freezing the catch made it available year-round.[5]

Local residents along the two rivers obtained their livelihoods from different sources, including crofting, reindeer farming, forestry, hunting, and berry picking. Along the northern rivers, fishing was a central source of food and income.

The nutritional value of salmon was significant to the locals, particularly along the lower parts of the rivers. Moving toward the upper reaches of the rivers, the amount of migratory fish decreased. At the mid-course of the rivers, salmon were still described as "half a livelihood," and hard-working fishermen in the upper parts of the river were able to bring home a nice addition to their dinner tables. Salmon had economic benefits as well, as migratory fish were often the sole source of spending money.

Salmon fishing required wide knowledge of the fish and the river, significant fishing skills, organization skills, and cooperation with other fishermen. Building weirs, seines, and traps also required special knowledge and skills. The amount and behavior of the salmon, as well as the features of the river at the fishing ground, dictated the fishing style that was used. Different methods were used in rapids and in quiet waters. The current, depth of water, and riverbed conditions at the fishing ground determined which type of trap was used.[6]

Fishing traditions were important builders of local and collective identity, as well as a significant part of the fishermen's personal identities, i.e., who I am and where I belong. Local people had grown up within a salmon fishing culture, and the traditions passed on from one generation to the next impacted their sense of belonging. Migratory fish were an important part of the family histories of many interviewees. There is a strong social aspect of salmon fishing, and seine fishing in particular was connected with the communal and social facets. The chain of generations shaped these identities; fishing was taught to small children, and they in turn passed on the tradition to their children. Fishing was seen as a bloodline, and continuing salmon fishing was often considered a matter of honor.[7]

Migratory fish also had an aesthetic value. A wanderer on the riverbank could appreciate the beauty and diversity of the river environment, and a salmon jumping in the rapids was a magnificent sight. Fishing was exciting; the thrill of salmon fishing was even greater because of their large size. If one did not have a fishing permit, the thrill rose not only from competition with the fish, but also with the fishing supervisors. Fish poaching was common, and the authorities often turned a blind eye to it. Torch fishing, in particular, is remembered by the interviewees as an exciting, phenomenal fishing experience. It is an old method of catching fish at night with a fish spear and a bright torchlight. The combined aesthetics of a dark autumn night, the beauty of water and fire, and the excitement of getting caught (torch fishing was forbidden) caused many interviewees to talk about nighttime fishing as an unforgettable memory.

Production of hydroelectricity in the Kemijoki started in 1948, right after the Second World War, with the construction of the Isohaara Dam and power plant at the mouth of the river. Harnessing the Kemijoki was one of the largest hydropower construction projects in Europe. Since then, a total of 17 large hydroelectric plants and two large water reservoirs have been constructed in the Kemijoki.[8] Harnessing the Iijoki started with the construction of the Pahkakoski Power Plant in 1959. In 1971, the last of the five power plants (at Raasakka) was completed.[9]

The damming of both rivers ended both the salmon migration and the rich fishing culture surrounding the migratory fish. When the interviewees talked about the end of salmon migration, they discussed the bitterness and discontent felt by local people. The biggest reasons for their discontent were loss of livelihood, loss of an important and enjoyable activity, and loss of the fishing culture. In a blind belief in the power of technology and its ability to solve any problems, functioning fish passages were neither planned nor built. When the power of Finland's northern rivers began to be harnessed in the 1940s, the production of electricity was generally seen as a common nationwide goal because the country was suffering from a severe energy shortage. Electricity was needed for the post-war reconstruction and war indemnity work. Those who endured losses due to the damming of rivers had neither any chance to challenge this goal nor any way to articulate their interests and express criticism of the prevailing economic and political situation. During the construction work, concern for social and environmental changes in Finland was minor; the typical mindset was that losses could be casually compensated by money. In the worst cases, people who had no experience with monetary calculations lost both their homes and their livelihoods. This was the case especially for the Lokka and Porttipahta reservoirs along the Kemijoki.[10]

Research Data and Methods

This chapter is based on ethnographic fieldwork conducted in 2009–2010, and in particular interview data, which consists of interviews with 23 salmon fishermen living along and near the Kemijoki and Iijoki rivers in northern Finland. The interviewees were elderly people aged between 60 and 91 years; interviews were carried out 60 years after the building of the Isohaara Dam in Kemijoki and 50 years after the building of the Pahkakoski Dam in Iijoki. The interviewees talked about their experiences fishing migratory fish, the activities that fishing and the use of fish included, and the changes they experienced in the river environment.

Thirteen of the interviewees lived beside the Kemijoki, while 10 lived beside the Iijoki. Six were female and 17 were male. The interview material collected from those who lived beside the Iijoki is part of a wider social impact assessment (SIA) completed for the Migratory Fish Return to the River Ii river restoration project.[11] I searched for interviewees along the Kemijoki by asking the municipalities' home help service for contacts and using snowball sampling.[12] The interviews were recorded on audiotape and then transcribed verbatim, and the analysis was based on content analysis.[13] I then analyzed the interviewees' perceptions of salmon agencies by interpreting the ways encounters with salmon were described, which features were related to salmon, and for what reason.

At this stage of the analysis, I avoided a tight framing of the concept of agency and instead looked for aspects of agency emerging from the interview material. However, some framing of the concept was needed to better guide the analysis. Agency is tied to subjectivity, even though the concept describes more than just an attribute of an individual. Related to subjectivity, agency is also inseparable from the surrounding social world. The concept of agency should be defined in relation to the production of subjectivity (as well as the definition of subject position) taking place in social processes. Thus, when defining agencies, power relations should also be considered, as they are present in all social relations.[14]

Offsets for Encounters with Salmon

Salmon were the biggest, strongest, and economically most significant fish, even though whitefish were even more plentiful in some locations. Other fish species, including whitefish, perch, pike, and grayling, were less valued. According to fishermen along the Iijoki, if you caught a whitefish, for example, people said the fish was bringing greetings from the salmon. Based on these greetings, the fishermen knew that salmon were coming soon and thus continued fishing.[15]

The interviews revealed three different standpoints regarding the human–salmon relationship: salmon as a superior, godlike entity, salmon as a commodity, and salmon and humans as equals. These perceptions are overlapping and interconnected with each other. Each one emphasizes particular aspects of the meaning of salmon and its role in the community (Table 13.1).

Salmon in Traditional Beliefs

Before the ideology of man ruling nature became concrete in these river areas—in this case, the construction of the first hydropower dams—the relationship between nature and people was different. Nature was considered stronger than human beings; local culture was dependent on natural conditions, especially with regard to the rivers. Nature was seen as a provider of life, and it was also given a spiritual and divine provenance. Man did not place himself above nature, but considered himself a part of the equilibrium of nature. In Finnish fiction, rivers are often described as streams of life and used as metaphors for life. The appreciation for rivers included spiritual and religious features.[16] Fish were seen as mysterious, supernatural, strong actors.

The gap between the primary elements of water, earth, and air and the unfamiliarity of water contributed to these philosophic beliefs. Water as the natural habitat of fish automatically segregates salmon from our own element. The underwater world is another dimension that is not reachable and remains essentially unknown to us. It was thus difficult to fully understand the attributes, environment, and

Table 13.1 Standpoints for salmon agency

Perception of salmon	*Subject positions*	*Emphasized meaning of salmon*
A godlike entity	Man as a part of the equilibrium of nature	Symbolic: metaphors Spiritual: animism Aesthetic: the diversity and beauty of nature
A commodity	Man above nature	Economic: nutrition, livelihood Social and aesthetic: recreation
An equal	Man and salmon as equal actors	Social: salmon as a companion Symbolic: salmon reflecting one's own life

behavior of fish. Due to the lack of a shared element and other known features of the fish, salmon were considered a mysterious creature, sometimes believed to possess supernatural powers. There is a connection between salmon and traditional beliefs and spirituality, as well as animism, as defined by Tim Ingold. For Ingold, animism requires more than just a simple sprinkling of agency. Things are in life, rather than life being in things.[17] Salmon, too, were believed to possess more than the given attributes of agency. Deidre Cullon studied the ceremonial uses of salmon along the Pacific Northwest Coast. She argued that an animistic belief endowed salmon with human qualities and agency. Proper care and respect ensured that these supernatural beings would remain benevolent and generous.[18]

Along the Kemijoki, the salmon fishermen were careful not to mention the name of the salmon while fishing. They believed that the fish could hear them speaking, and when recognizing its own name, the salmon would hide. Thus, many alias names, which were often borrowed from other languages, were used for salmon. When an alias name became common, it had to be changed again.[19] Besides the possession of agency, the variety of different names for salmon also speaks to its importance in the culture of riverside communities. Wooden or stone sacrificial landmarks for fishermen's luck are signs of spirituality and animism and they can still be found around fishing grounds in northern Finland.

The next example reveals some aspects of the supernatural beliefs surrounding salmon, as well as the fact that men and salmon were perceived as equals. Veikko, an elderly man who used to fish for salmon along the Kemijoki, told a story about his father and grandfather. His grandfather had gone away for the night and he told his son, the narrator's father, to examine the salmon net the next morning. The son did as he was told and caught two salmon. Because he had been short of money, the son decided to sell the second salmon and not tell his father about it. When his father came home, he immediately asked about the salmon net. "There was one salmon," his son told him. "Not two?" marveled the father. "There should have been two. Last night in my dream I was fighting two men."

Salmon as a Commodity

Opposite perspectives of the human–animal relationship are obvious in the data as well. Especially downstream of both rivers, salmon were severely overfished. Some of the interviewees described the salmon culture and fishing in the same way they talked about haymaking or harvesting potatoes. Many animals are affected by their location within a distribution of resources that is deeply skewed toward humans. Their habitats and ecologies are subject to human interests and depredation; their lives are subordinated to the needs of humans; and the material and ecological conditions for their survival are frequently disregarded during the pursuit of human commercial advantage.[20] Some interviewees highlighted the nutritional and economic value of fishing, describing their methods and tools, but also spoke about the social and recreational aspects of fishing, where the river environment and the element of flowing water were highly important. In these narratives, human activity and agency were emphasized and salmon were seen as an object, a commodity without any agency of their own. In the next interview excerpt, Ilmari from the Kemijoki River talks about the annual rhythm of salmon fishing and compares the different traps:

ILMARI: When the river was still free, it was the beginning of July when the salmon began to appear. One could not immediately catch them with lures, but later in August they were caught with nets and lures, and all autumn until the river started to freeze. Some salmon stayed, which were called "winter salmon," and they were fished already early in the spring near Juoksuniemi, our cemetery hill. Seine was a good trap, if you were able to throw it wider. My father used to fish with lures; he got many salmon, but the nets, they were not so expensive.

Reading between the lines, however, one could reveal the hidden agency of the fish as respect for a strong fish and a provider of life. The different perspectives of salmon seem to be overlapping, and the perception which dominates the others depends on the context of the telling.

Salmon as a Mirror for Human Agency

According to the third perspective, salmon were considered similar to humans. The perceived equality between humans and salmon may have partly been due to the large size of the fish. The weight of one fish could reach almost 30 kilos, although most were 10–20 kilos. Fishing such specimens required both skill and strength. A competitive spirit was present in the interview data in two different forms. The battle was perceived on the one hand as taking place between the fisherman and the catch, with the fishing event described as a battle between two equal partners. It was also a battle between man and nature, as survival was at stake. On the other hand, the competition was between the fishermen: who would get to the best fishing grounds, who would get the most salmon, and who would get the biggest fish. Catching a large salmon was a sign of skill, strength, and cunning. In a way, it strengthened the fisherman's status within his community; the prize in this competition was the appreciation of others and a significant role in many fishing stories.

Two interviewees, Alpo and Hannes, described wrestling with salmon:

ALPO: It was customary to jump over the salmon when it was caught in shallow water, because of the fear of losing the catch. This was possible because the salmon were not as strong in shallow waters. When wrestling with the fish, one could feel how strong it was. If you held it between your knees and grabbed its gills, it could still get away if its tail was free.

HANNES: Yes, if the salmon's tail was free, one could not hold it back.

The interviewees saw similarity between the salmon lifecycle and their own behavior, appreciations, and intentions. It was well known among the fishermen that salmon migrated in shoals of fish, and this was considered a sign of their social character. Salmon, like people, were sociable creatures that appreciated the company of others. They belonged to a collective of salmon and were members of a family. Sometimes the fish also seemed to search for contact with the fishermen, as illustrated in Niilo's interview.

> NIILO: We were going to the Hepokangas fishing ground when a female salmon made an appearance. It was so cunning, one had to think it was looking for company, and as it heard some clatter, it showed itself swimming right near the surface of the water. It did not leap; it just showed its dorsal fin. Aato said, "You just wait for us, we will be here with the torches tonight." So we went, and the stupid salmon came swimming along to see what we were up to, and Aato caught it right away.

Many interviewees perceived salmon to be cunning, often even playing with the fishermen and teasing them. One day, after hours of fishing with no success, Pekka decided to give up. He rowed his boat to the shore, walked to his cottage located on the riverbank, and slammed the door behind him. Immediately after this, a big salmon leapt right next to Pekka's rowing boat. The same thing happened three days in a row. Pekka saw this as a sign of malicious pleasure on the part of the salmon, but also as an invitation to play a bit more.

Some thought that salmon, similar to people, used to search for cool places during hot weather. When the weather was hot, the fishermen searched for salmon in cooler deep waters. When salmon were caught, the fishermen often saw signs of astonishment in the fish. Still, even if they had a small chance to escape, the salmon did not always seize the opportunity. This, as interpreted by the interviewees, was due to their amazement – the fish were so confused and surprised that they momentarily had no idea where to go.

Laziness was another characteristic of salmon that interviewees mentioned. This was based on the observation that salmon were often seen at rest in still waters, especially after breeding. These "lazy" salmon were rather easy to catch with a fishing net or gig. In the local culture, hard work and diligence were greatly appreciated, especially among the older generation. The interviewees mentioned hard work as an important value and a source of contentment in their lives; they started working very early, and hard work had been constant ever since. Their traditional social system leaned much on Lutheran philosophies, which emphasized the value of a strong work ethic. Diligence increased one's social appreciation, belonging, and agency within the community, while laziness and an idle attitude toward work were not approved. From this viewpoint, the resting salmon were not appreciated as much as the ones that leapt for their breeding areas.

The fact that salmon return annually to their birth river to breed was well known among the people along the rivers. It was considered a deeply human characteristic; fish were perceived to express appreciation for their birthplace. The river was considered the "right" place to be – fish came "home."

ARI: Salmon are born here and that is significant. If they are born on a fish farm, it is not the same thing. It is not their home. I myself know that my home is along the Iijoki [laughs], and so do the salmon. If they have a chance to return, say to some rock in Livojoki [tributary of Iijoki], they will go on to their trip to the sea and then come back. They have the tendency to do so. And they return to exactly the same place where they were once born. We think we know much, but salmon are wiser. We do not have the kind of direction system that salmon have. Even if we had a GPS, salmon return to precisely the same place without a GPS, and their trips are thousands of kilometers away.

According to Carter and Charles,[21] the natal circumstances of all human beings place them within historically specific and determinate social relations and arrangements. In the interviewees' minds, the fact that salmon wanted to raise their offspring in the same place they were born was connected to people's own place attachments and appreciation for their family roots.

Evidence of social hierarchies and dominance were also expressed. Some elderly men saw a difference between female and male salmon.

NIILO: The male salmon could not be caught. Were they blessed or what, damn! Or maybe they were so much wiser that they did not get caught, never. It was always just females. And I have never seen a female salmon bigger than 14 kilos. Even that is out of the ordinary. Females are usually 7 or 8 kilos, more or less than 10 kilos. But we could not catch male salmon, except by torch fishing, and sometimes with seine.

Female salmon were considered stupid and were not appreciated as much as males, especially large, awe-inspiring male salmon. Males were considered more cunning, wiser, and much more difficult to catch. Male salmon were more active than females, and when breeding, males were thought to actively look for and choose the breeding place. Niilo used an interesting word, "*viihteenuinti*," or "entertainment swim," for salmon's breeding time. He explained that during the entertainment swim, it is usually the male salmon that swims around and looks for a partner, searching through all the breeding areas. Male salmon were also considered smart enough to avoid capture. They teased the fishermen by showing their tail or dorsal fins on the water surface. If such a cunning male salmon happened to be caught in seine fishing net, it understood to keep quiet until the last moment.

Many interviewees talked about having emotional bonds with the fish and expressed caring and empathy for them. This tells about an equal approach. The fact that dam construction ended the salmon migration was painful for many interviewees; they grieved for the fish. For example, Eeva, who had seen the salmon's attempts to overcome the dams, said that during her long life she could never forget or forgive the painful memories of salmon suffering. She witnessed how strong the salmon's intentions were when they seriously injured themselves by leaping against the dams.

EEVA: And that summer Isohaara [the dam] was closed and listen to me, the injured salmon tried to leap in between the logjam and timber … [her voice fades] it was just horrible. They could not pass. I have seen such a sad thing there and it has stayed in my heart as a bad memory: that they can do that, just close the best river in Finland.

Along the Iijoki, a similar story was told by Alpo. He witnessed the final construction work at Pahkakoski Dam, and after the hatchway of the dam was closed, a shoal of salmon started leaping against the dam constructions. Alpo was impressed by and convinced of the persistence of the fish, which he considered another sign of salmon agency.

ALPO: They leapt there; you know a salmon never gives up. They try and try, they leapt against the stream, and the stream hit them back against the breakwater made of concrete. Masses of salmon got injured because of these breakwaters. You could see them, the ones that were physically injured.

After seeing the drive that forced salmon to leap against the dams, the fish were caught at Pahkakoski Dam and transferred past the hydropower plant. The fish, however, seemed to have lost the instinct to continue swimming toward their breeding habitats. People assumed the fish suffered from stress and placed the salmon in special stress pools for 24 hours, which was believed to calm them down. Next, the salmon were taken back to the river, past the dam. However, these attempts did not work and they could not save the salmon stock; this procedure was gradually renounced.

Interviewees also worry about the changed circumstances in fish habitats.

EEVA: And there at Jyrhämä the shores and riverbanks were modified because of the Vanttaus Dam, because the dam raises the water level. I think it is wrong; they do not like the fact that the shores were modified with big rocks and stones. This is my own…

ILMARI [INTERRUPTING]: I think it is good that the hydropower plant has been built, it has improved the economy. The shores should not disturb anyone – at least they do not disturb me.

EEVA: No, but I mean the fish, because the fish cannot get to natural shore anymore, just to the rocks. Here at Jyrhämä, I mean. But I might be wrong.

Eeva and Ilmari shed light on questions like "Who has the right to talk?" and "Who has the right to the river?" In Eeva's comments, the destinies of fish and riverside people seem similar. Facing prevailing economic and political pressure, both entities must quietly accept their fate. People are discontented with the environmental change in the river landscape, but they do not dare to speak up and express these feelings. The dominant narrative of the change in the river is based on legal, technical, and economic language; against the good of society, pleading for one's own experiences is difficult and useless. Eeva knows that her opinions are too modest and have no room in the discourse on hydropower, energy production, or the state economy. Thus, she appeals for the well-being of the fish; the fish do not like the fact that the shores have been altered. She places fish first and talks about the losses fish have experienced rather than her own losses. Ilmari refers only to economic rationalities, considering Eeva's comments incorrect and foolish.

During the last decade, several migrant fish restoration projects have taken place on these rivers. The interviewees are still concerned about the well-being of salmon. They know about these river restoration projects through the media and follow the projects with interest. One restoration activity involves catching salmon from the river mouth and transporting them to the upper parts of the river, past the hydropower plants. The interviewees worry about how the salmon are coping with these stressful situations. They discuss how salmon should be looked after, and hope that the turbines of hydropower plants will not hurt fish on their way back to the sea.

AINO: But now they worry about the salmon that they planted in Ounasjoki [a tributary of the Kemijoki].

OUTI: Yes, they transported salmon there.

AINO: Yes, now they talk about how a salmon has escaped and if they have found it or ask if someone caught it [laughs]. I have read in the newspaper, and heard discussions on the radio too, about how many smolts there will be next year. I have laughed many times with my son and wondered who will observe the salmon day and night on the riverbanks.

Conclusions: Intertwining Agencies

The northern sources of livelihood and the factors that people living in these challenging circumstances depend on, such as salmon, shape the local cultures in many ways. The economic significance of salmon fishing also creates and reconstructs meanings in other sectors of life, including the social and spiritual sectors. Many interviewees saw salmon as a strong actor. Symbolism, as well as humanizing and caring for the fish, was common in my data. According to Zelko, attitudes toward specific species are largely shaped by our innate anthropomorphism.[22] That is, when we think about animals, we are also thinking about ourselves. Additionally, some interviewees talked about the agency of the river. The river was also considered an entity with thoughts, actions, and objectives. This came up when the interviewees described the change from a free-floating river to a heavily regulated one. A free river's desire is to reach the sea, flowing freely toward its destination, and building a dam is a deathblow. This metaphor also fits the salmon, who share the same sad destiny. Salmon fishing involved an interaction between three partners: the fisherman, the salmon, and the river (animal–human–environment interaction). Some interviewees added traps as a fourth component. This became especially common in narratives about fly-fishing.

The role of the fish in local riverside cultures was much more than just economic. The interviewees have grown up in a fishing culture filled with stories of salmon; they observed and learned about salmon behavior while fishing, and interpreted and compared the stories and observations with their own life.

KALEVI: Indeed, my father was a fisherman, and he took me along to the river from a very early age. Since I started fishing as a little boy, fishing for all kinds of fish in the river, of course it went all the way to my blood; you couldn't help it.

Narratives about the qualities of the fish, as well as the descriptions of fishing activities, showed that salmon were believed to have agency. My data include a variety of salmon agencies with different intensities. At a minimum, salmon were perceived as a commodity, and the human–animal relationship resembled an actor–object relationship. The agency of fish in this composition was narrow. However, some interviewees gave salmon a much wider agency as a companion, human-like creature mirroring the lives of riverside people. Salmon characterizations and comparisons were drawn from the interviewees' own familiar everyday environment, and in these narratives human and salmon agencies were intertwined. Moreover, according to traditional beliefs and spirituality, salmon were seen as a godlike creature possessing supernatural powers.

The actions of the fish were regarded as evidence that salmon possess agency. The fish were seen as having the ability to take action and as possessing thoughts, emotions, and intentions. Things that were known about the fish were reflected in human life and interpreted and explained by comparing them with one's own experiences and intentions. Salmon were enmeshed within social and cultural relationships. In this way, local people along the riversides brought salmon into a social world that was familiar to them. Salmon agency was relational. As Carter and Charles have written, in this sense, non-human animals are agents in relation to human-dominated structures.[23]

The river and its damming put people and salmon in the same restricted position, as both depended on a free river. People were aware of what they shared in common with the salmon. When the electrification of the rivers began, local residents saw the sad fate of salmon and were reminded of their own situation. Reflecting one's own experiences was now reciprocal.

Defining agency is a social process. As linguistic creatures, people define the agencies of salmon as well as the extent and dimensions of such agencies. All of these agencies are produced in interaction, which in turn is always contextual, time-, action-, and place-specific. Agency is negotiable, and it is strongly related to power relations. When defining salmon agencies, subjects, wider subject positions, and power relations are produced and reproduced at the same time. As interviewees talked about salmon, they defined not only salmon agencies, but also positioned themselves and described the surrounding community.

Notes

1 Autti, *Valtavirta muutoksessa*; Autti and Karjalainen, "The Point of No Return," 45–57.
2 Vilkuna, *Lohi*.
3 Ibid.
4 Hoffman, *Pohjolan Voima 1943–1993*.
5 Vilkuna, *Lohi*.
6 Vilkuna, *Lohi*; Autti, *Valtavirta muutoksessa*; Autti and Karjalainen, "The Point of No Return."
7 Autti, *Valtavirta muutoksessa*; Autti and Karjalainen, "The Point of No Return."
8 Suopajärvi, *Vuotos- ja Ounasjokikamppailujen kentät*; Kemijoki Oy, "Voimalaitokset ja tuotanto."
9 Rusanen, *Role of the Local People*.
10 Autti, *Valtavirta muutoksessa*; Järvikoski, *Vesien säännöstely ja paikallisyhteisö*; Luostarinen, *A Social Geography of Hydro-Electric Power Projects*.
11 Karjalainen et al. *Monitavoitearviointi Iijoen vaelluskalakantojen palauttamisen tukena*.
12 For example, Noy, "Sampling Knowledge: The Hermeneutics of Snowball Sampling in Qualitative Research," 327–344.
13 Neuendorf, *The Content Analysis Guidebook*.
14 Hughes, *Key Concepts*; Ronkainen, "Haavoittunut kansakunta," 531–550; Virkki, *Vihan voima*.
15 Autti, *Valtavirta muutoksessa*; Autti and Karjalainen, "The Point of No Return."
16 For example, Pietarinen, "Ihminen ja luonto," 98–108; Vilkka, *Ympäristöetiikka*.
17 Ingold, *What Is an Animal?*
18 Cullon, "A View from the Watchman's Pole," 9–37.
19 Vilkuna, *Lohi*.
20 Ingold, *What Is an Animal?*; Carter and Charles, "Animals, Agency and Resistance," 331.
21 Carter and Charles, "Animals, Agency and Resistance," 331.
22 Zelko, "From Blubber and Baleen to Buddha of the Deep," 91–108.
23 Carter and Charles, "Animals, Agency and Resistance," 331.

Bibliography

Ethnographic Materials

All field journals, audio, and transcripts of the interviews are in the possession of the author.

Literature

Autti, Outi. *Valtavirta muutoksessa – vesivoima ja paikalliset asukkaat Kemijoella*. Oulu, Finland: Oulun yliopisto, 2013.

Autti, Outi, and Timo P. Karjalainen. "The Point of No Return: Losing Salmon in Two Northern Rivers." *Nordia Geographical Publications, NGP Yearbook 2012* 41 (2013): 45–57.

Carter, Bob, and Nickie Charles. "Animals, Agency and Resistance." *Journal for the Theory of Social Behaviour* 43 (2013): 322–340.

Cullon, Deidre. "A View from the Watchman's Pole: Salmon, Animism and the Kwakwaka'wakw Summer Ceremonial." *BC Studies* No. 177 (Spring 2013): 9–37.

Franklin, Adrian S. "Performing Acclimatisation: The Agency of Trout Fishing in Postcolonial Australia." *Ethnos* 76 (2011): 19–40.

Hoffman, Kai. *Pohjolan Voima 1943–1993*. Oulu, Finland: Kaleva, 1993.

Hughes, Christina. *Key Concepts in Feminist Theory and Research*. London: Sage, 2002.

Ingold, Tim, ed. *What Is an Animal?* London: Unwin Hyman, 1988.

Järvikoski, Timo. *Vesien säännöstely ja paikallisyhteisö*. Turku, Finland: Turun yliopisto, 1979.

Karjalainen, Timo, Anne-Mari Rytkönen, Mika Marttunen, Aki Mäki-Petäys, and Outi Autti. *Monitavoitearviointi Iijoen vaelluskalakantojen palauttamisen tukena*. Helsinki: Suomen ympäristökeskus, 2011.

Kemijoki Oy. "Voimalaitokset ja tuotanto." Accessed September 20, 2016. www.kemijoki.fi/toimintamme/voimalaitokset-ja-tuotanto.html.

Luostarinen, Matti. *A Social Geography of Hydro-Electric Power Projects in Northern Finland: Personal Spatial Identity in the Face of Environmental Changes*. Oulu, Finland: University of Oulu, 1982.

Neuendorf, Kimberly. *The Content Analysis Guidebook*. Thousand Oaks, CA: Sage, 2002.

Noy, Chaim. "Sampling Knowledge: The Hermeneutics of Snowball Sampling in Qualitative Research." *International Journal of Social Research Methodology* 11 (2008): 327–344.

Pietarinen, Juhani. "Ihminen ja luonto: neljä perusasennetta." In Arvot, hyveet ja tieto, edited by Pekka Elo and Hannu Simola, 98–108. Helsinki: Painatuskeskus, 1995.

Ronkainen, Suvi. "Haavoittunut kansakunta ja väkivallan toimijuus." In *Väkivalta – seuraamukset ja haavoittuvuus. Terttu Utriaisen juhlakirja*, edited by Mirva Lohiniva-Kerkelä, 531–550. Helsinki: Talentum, 2006.

Rusanen, Jarmo. *Role of the Local People in the Utilization of Water Resources: A Case Study of the River Iijoki in Northern Finland*. Oulu, Finland: Pohjois-Suomen maantieteellinen seura, 1989.

Suopajärvi, Leena. *Vuotos- ja Ounasjokikamppailujen kentät ja merkitykset Lapissa*. Rovaniemi, Finland: Lapin yliopisto, 2001.

Vilkka, Leena. *Ympäristöetiikka. Vastuu luonnosta, eläimistä ja tulevista sukupolvista*. Helsinki: Yliopistopaino, 1993.

Vilkuna, Kustaa. *Lohi. Kemijoen ja sen lähialueen lohenkalastuksen historia*. Keuruu, Finland: Otava, 1975.

Virkki, Tuija. *Vihan voima. Toimijuus ja muutos vihakertomuksissa*. Jyväskylä, Finland: Atena, 2004.

Zelko, Frank. "From Blubber and Baleen to Buddha of the Deep: The Rise of the Metaphysical Whale." *Society & Animals* 20 (2012): 91–108.

14 The Co-Living of Humans and Wolves in the Greater Yellowstone Ecosystem

J. Dwight Hines

Introduction

In this chapter I analyze the relationship of humans and wolves in the Greater Yellowstone Ecosystem in the late twentieth and early twenty-first centuries. Specifically, I show how the relationship reflects significant cultural shifts within the history of European-American occupation of the western United States. In the process, I will also highlight the role that animal agency plays in the creation of the contemporary human–wolf relationship.

Prior to European colonization of North America the relationship between gray wolves and humans was largely characterized by degrees of mutual respect. For the first four hundred years of European-American hegemony on the continent, however, the wolf was hunted mercilessly as an enemy of economy and society, culminating with the almost complete extinction of wolves in the United States by the middle of the twentieth century.

Shortly thereafter, however, there burgeoned an ethos of conservation among the American middle class, as part of which emerged the idea of reestablishing wolf populations in parts of the United States. Wolf reintroduction efforts in Yellowstone National Park (YNP) beginning in 1995 can be seen as a partial triumph of an environmental agenda, which is itself a harbinger of the rise of what I call the "experience economy." As I chronicle elsewhere, the "experience economy" is one of the latest elaborations within the evolving capitalist world-system centered on the rising emphasis on the production/consumption of "experiences" – that is, episodes within human life without necessary impact on material existence, which are themselves often the products of concerted and significant capital investment, programs of creative construction, and substantial systems of material production – in the lives of the ever-emerging post-industrial middle class.[1] The increasing efficacy of this agenda and economy has contributed to a simmering conflict within the United States over wolves, especially in the American West, which continues to this day.

Within a cultural climate of capitalism, species like the wolf represented a profound challenge to the ideas of success. The wolf, by preying upon the products (that is, livestock) of American agriculture, disrupted its efficiency and potential for economic viability. Such impediments were meant to be removed, permanently if possible, hence the centuries-long all-out and nearly successful war on them.

This cultural perspective and project began an about-face, however, in the middle of the twentieth century with the steady expansion of the post-industrial middle class within US society. It is amongst this subset of the overall middle class that you see an increasing emphasis upon the production and consumption of "experiences." For these people nature – those places described as rural or wilderness – is not to be used, first and foremost, to create commodities from silviculture, mining, and agriculture but rather to foster satisfying recreational and lifestyle practices, which can be marshaled and deployed as symbols (that is, markers of social distinction) within an emerging post-industrial language of social difference.

Part of the thesis of this chapter is that the ideological and rhetorical attention given by the American middle class in support of the reintroduction and conservation management of wolves is indicative of their subscription to a post-industrial regime of the production/consumption of experiences. I also propose that it is the industrial regime of production/consumption which is largely the ideological and rhetorical province of the American working class. For them it is the production and consumption of "things," such as livestock, timber and minerals, that they believe constitutes the "highest and best" use for the lands of the rural American West; anything – a being, a policy, etc. – which actually or potentially interferes with that program, such as predatory animals, especially ones with endangered species status, are considered to be contradicting the intended purpose of the land. Thus, animals, such as the wolf, are drawn into human political conflicts as ciphers or symbols; such animals, however, must be understood as sentient beings with degrees of agency as well.

With regards to the second portion of this chapter, it is perhaps necessary to say a few introductory remarks as to how I conceive of the concept of animal agency. A premise of this analysis is that the interpretation of behavior or the intention behind behavior by humans is influenced by the patterning of learned and shared perspectives (what anthropologists call culture). To put it another way, our experience of everything around us is filtered through and therefore subject to the meaning we give to it. The same is true for how we look at physical processes of the earth – for example, weather or plate tectonics – or animal behavior. Due to the limits of consciousness none of us can ever demonstrably know why another human has behaved in a certain way. We can and do infer, based on a myriad of connections we can draw (based first and foremost on our shared humanity and then based on degrees of shared cultural perspectives – to the extent that they exist), why we think another human has acted as s/he did. Despite these connections, the project is still yet predicated on supposition, informed supposition, but supposition nevertheless. The same is obviously true when it comes to our attempts to interpret other animals' behavior, only in these cases we have even less connections upon which to draw for we do not have a discernibly shared consciousness, let alone shared degrees of culture, to draw upon. All this is simply to say that interpreting behavior is a complex project in this world and it is always open-ended and incomplete between humans and all the more so when we try to describe agency in other animals. That being the case, however, does not absolve us of trying. Thus I will conclude this chapter by suggesting a provisional interpretation of the role that agency plays in the creation of the relationship between humans and wolves in the Greater Yellowstone Ecosystem (GYE).

A Brief History of Wolves in the Greater Yellowstone Ecosystem

The gray wolf (*Canis lupus*) is an indigenous species of North America whose range encompassed most of the continent prior to European colonization. Extensive hunting throughout the frontier period both as a means to prevent predation of livestock and out of the deep-seated cultural biases against the animal led to the radical reduction of its numbers and range. By the twentieth century the wolf survived in only a few pockets in the United States, most notably in mountainous regions of the American West. By the mid-1930s the last wolves were gone from YNP.[2]

In the 1960s came the call for the reintroduction of the wolf to the wild lands of the Rocky Mountains. As early as 1968 the US National Park Service expressed recognition of its role in the production of ecosystems and its desire to deal with the missing components thereof.[3] Despite these nascent rumblings, regional sentiment as well as local political will was still significantly opposed to the effort.

The late 1970s saw a gradual development of the wolf reintroduction idea. In 1978, US National Park Service biologist John Weaver proposed the specific idea to "transplant wolves [from Canada or Minnesota] … to restore a viable population of this predator" to YNP. In 1980 an initial plan for reintroduction was drafted and a revised version was signed in 1987. By the late 1980s, popular sentiment had also shifted; at the time, a survey of YNP visitors indicated overwhelming support for wolf reestablishment.[4]

In 1990, the US Congress appropriated $375,000 to support the development of a wolf introduction and management plan for YNP. In 1995 and 1996, the National Park Service succeeded, in conjunction with the US Fish and Wildlife Service, in introducing 66 individuals, which were tranquilized, captured and transported from Alberta, Canada, into YNP.[5] The numbers of animals rose so rapidly that by 2000 the numbers had already reached the initial Fish and Wildlife Service goals of 300 individuals and 30 packs in YNP. Since then the numbers have continued to rise with the most recent 2010 Rocky Mountain Wolf Recovery Interagency Annual Report indicating that there were at least 1,651 individuals in 244 packs with 111 breeding pairs in Montana, Wyoming, and Idaho.

While the numbers overall in the northern Rocky Mountain states have gone up, the numbers in YNP have fallen from a high of 94 in 1997 to fewer than 40 in 2010. Biologists hypothesize that this ironically is a symptom of the wolves' success as they have effectively culled the park's ungulate herds, especially elk. As their sources of easy prey have dwindled, wolves have been forced to move outside their federally protected domain. This point is bolstered by the numbers, which indicate a precipitous drop in the numbers of elk in YNP, particularly in the so-called northern herd, which saw its population decline significantly since wolf reintroduction. This circumstance, beyond inspiring more wolves to move outside the park, has incurred the ire of local professional outfitters, who rely upon the availability of such big game to sell to out-of-state hunters. Although the numbers of the northern and southern Yellowstone elk herds have declined since wolf reintroduction their numbers are still well above the goals set by the US Fish and Wildlife Service.[6]

The ongoing discussion of wolves in Wyoming, Idaho, and Montana some 17 years after reintroduction is not something people on the conservation side of the debate could have anticipated. In many respects the wolf is now the potential victim of its own success. As noted above, the numbers have steadily risen and wolves have expanded their territory beyond YNP. In so doing, they have left the confines of the National Forest Service lands and even the GYE, which in turn has thrown them into further conflict with the residents – especially ranchers and hunter – of the surrounding states. This has shifted the issue from one of whether or not the wolves should be brought back to how they should be managed.

From the anti-wolf side, the expansion of the canine's range beyond public lands confirmed their worst-case scenario: the wolves were uncontrollable and they would reoccupy significant swathes of the Rocky Mountain West. Increasingly these people called for the delisting of the wolf as a federally mandated endangered species and for management of the animals to be turned back to the individual states. In 2009, the US Fish and Wildlife Service agreed to delist the animal and to give control back to the states of Idaho and Montana, which had drafted management plans that met the basic guidelines – of 150 wolves and 15 breeding-pairs – thought necessary for population maintenance. However, the Fish and Wildlife Service balked at extending the same opportunity to Wyoming since that state persisted in its proposal to designate the wolves as predators outside of the GYE and allow for the unregulated killing of them.

Crying Wolf: The Divergent Perspectives on the Value of Wolves in Yellowstone

Over the last decade of my research in the northern Rocky Mountains,[7] it has become evident that wolf management is one of several issues that glaringly serve as ciphers for the emerging class-cultural contest between post-industrial middle-class and industrial working-/middle-class Americans within the "New" West Archipelago, which is a term for the string of "island" communities of the mountainous American West that are facing profound change from the in-migration of significant numbers of former urbanites in the last three decades.[8]

When addressing the topic shortly after reintroduction of wolves to YNP, Wilson identified three principal issues as driving the debate between the members of two social movements – the environmentalist and the so-called "Wise Use."[9] There were: "(1) differential access to social power, (2) conflicting ideas about private property, and (3) divergent beliefs about humankind's proper relationship with the natural environment."[10]

To those I add a fourth issue, which springs from the emerging difference between the two competing ideals of proper land use, especially the "public domain," that is, the federally administered lands that make up over half of the western United States. This difference, furthermore, is exacerbated by the expanding subset of the US society that values "experiences" as part of their post-industrial regime of production/consumption.

The Industrial and the Post-Industrial Perspective

Strictly speaking, there is nothing new about the social subset upon which this chapter focuses. Throughout the modern capitalist era there has existed a segment of Western society that matches the profile of the post-industrial middle class to be described here. Its initial incarnation was practiced by the scions of the early English bourgeoisie, those who knew the value of and could afford "The Grand Tour" of the art and architecture of the continental European capitals as a sign of their distinction from the uncultured masses. In this sense, this nascent form of tourism was the enactment of the traveler's distinguishing levels of finance and education that empowered him/her to pursue and appreciate proper experiences. Over the course of time, as Western societies mutated under the pressure of a dynamic capitalist modernity, the post-industrial middle class – the social subset that relies upon experiences as a significant marker of social distinction – has slowly but steadily expanded both in number and relative prominence. This expansion accelerated with the marked increase in post–Second World War affluence and the swelling of the occupational ranks dedicated to professional and managerial services, as well as education and other "creative" pursuits (of knowledge, art, entertainment, etc.).

In the context of the Montana community where I work, the exurbanites invariably come from (sub)urban middle- or upper-middle-class backgrounds, are college-educated, and bring, create, and encounter occupations in the professional or "service" industries. They are economists, lawyers, artists, and baristas, to name a few examples. They are exactly the people who are engaged in the creation, manipulation, and development of images and symbols and, therefore, are especially predisposed to be the primary and most ardent consumers of the same experiences.

As easy as this may seem, it is important when analyzing such middle-class formations to "avoid simply reading different groups off from their 'objective' class/occupational positions."[11] While position within the middle class is largely predicated on socioeconomic factors, such as profession and income, the post-industrial middle-class power and position are "always exercised and reproduced culturally";[12] its members must "act right" in order to be accepted as members of the class or class faction. Thus, as with all social characteristics, class must be understood as a performance or set of practices. Increasingly "acting right" for the post-industrial middle class is judged by the proper production and consumption of experiences as well as the production and consumption of proper experiences, which differentiate its members from co-existing subsets of the middle class.

From this we can appreciate that post-industrialization, in general, does not necessarily entail a greater emphasis on consumption over production. Instead, as I describe it here, it involves the shift from the production and consumption of commodities to the production and consumption of experiences. Thus, members of the post-industrial middle class are enculturated to a regime that prioritizes the production and consumption of experiences, such as knowledge, services, and entertainment, as marks of their distinction from the members of other subsets of US society.

These differing perspectives on what constitutes appropriate human behavior translate readily into ideals about the prescribed way in which to relate with and manage the land – and the animals thereof. Thus, these competing images of proper personhood translate to other animals as well. We can decipher in the differing attitudes toward contentious animal issues, such as those surrounding wolves in the GYE, the competing cultural perspectives of the combatants.

The Industrial Perspective on Wolves

If you look intently, as I have through my ethnographic fieldwork, you find that those residents of the Western United States who express a bias against wolf reintroduction tend to posit the argument that the animal is a threat to the real purpose to which the land, especially the public lands of the American West, ought to be dedicated – what is pithily referred to in local parlance as the "highest and best use." You also find that these folk – a version of industrial Americans – believe that the public lands of the West are best used primarily, if not exclusively, for the production of agricultural goods, minerals, timber, and big game wildlife as saleable commodities.

With the advent of state-level management in Idaho and Montana, local anti-wolf activists have been emboldened. This led to a proliferation of propositions on how to improve the agreements made between the states and the federal government. Both states have seen proposals by anti-wolf advocacy groups to alter the agreements to permit more and different ways to hunt wolves. These movements have each attempted to recast the situation as a "takings" issue, which is a common rhetorical and ideological stance adopted in such contests. By proposing that the loss of livestock or wildlife represents a confiscation of private property by the federal or state governments, wolf opponents seek to tap into the rich vein of libertarian leanings and resentment for "outside" government influence in the rural American West.

For example, in western Montana a group of livestock producers are pressuring the Ravalli County Commission to adopt a system to reward people who kill predators in the Bitterroot Valley. If implemented, the plan, which was proposed by the Livestock Protection Group, would fund a $100 bounty on wolves and mountain lions. It is unlikely that any such measure by a county would be able to supersede the jurisdiction of the Montana Fish, Wildlife and Parks on such matters.[13] Nevertheless, the attempt itself speaks to the extent to which the anti-wolf position has been invigorated by recent developments.

At the same time, just north in Missoula County, anti-wolf advocates have proposed a similar plan. Toby Bridges, the self-professed voice of a group of "sportsmen against wolves" has accused the Montana Fish, Wildlife and Parks of "working closely with radical environmental groups, the majority of which are extremely anti-hunting."[14] One of these groups, the Yellowstone to Yukon Conservation Initiative, Bridges said, is dedicated to "make it where you can't live on the land" and to "reduce human use of the West."[15]

The Post-Industrial Perspective on Wolves

Within the pro-wolf camp there are a range of related positions. One extreme is evident in the discourse of the Greater Yellowstone Coalition. The Greater Yellowstone Coalition was formed in 1983 as a confederation of 60-plus independent non-profit organizations. It is dedicated to keeping the GYE "healthy and wild" by keeping it "whole."[16] The Greater Yellowstone Coalition's flagship programs involve predator advocacy: grizzly bears and wolves are the primary species upon which they focus. Its stated values regarding wolves are three-fold:

1 Wolves are critical to the overall health of the Greater Yellowstone Ecosystem as a keystone species, restoring ecological balance. Most notably wolves improve the health of their prey through selecting weak, old, diseased and injured animals.
2 Wolves also change the habits of prey, such as elk, by reducing their numbers and changing distribution. These changes have contributed to a rebirth in the growth of cottonwood, willow, aspen and shrubs, subsequently benefiting grizzly bears, pronghorn, beaver, cutthroat trout, songbirds, scavengers and small mammals.
3 Wolves have proven to be a substantial economic draw for Yellowstone and Grand Teton gateway communities, with an estimated $35.5 million annually attributed directly to visitors coming to the three-state region solely to see wolves. In Yellowstone, almost half of the more than 3 million annual visitors are specifically interested in the possibility of seeing a wolf. Eco-tourism has developed as a valuable business in communities surrounding Yellowstone.

It is important to note that all three of these values are ultimately predicated on the intangible experience they provide to us as human beings, tourists, and people concerned with the United States' status as a leader in conservation efforts.

A more moderate pro-wolf position is heard in the voices of people living in the gentrifying "New" West Archipelago.[17] This position is characterized by a resignation to the fact of the wolves' presence in the northern Rocky Mountains. This resignation comes with an appreciation that the future management of the wolves will demand a significant compromise from both extremes of the debate.

Evidence of some acceptance of this situation from people on the industrial side of the divide comes in the voice of Mr. Peterson, a fourth-generation rancher in Montana's Big Hole, who says, "I do not dislike or hate the animal" and adds that "the wolf is here to stay now, and my feeling is that those people who want it here should share the costs."[18]

A similar perspective to Peterson's comes from Davis Reagan, a recent in-migrant:

> As far as I can tell, I'm right in the middle when it comes to having wolves here ... I support their reintroduction and am glad they are here again. They are beautiful animals and efficient predators with truly impressive intelligence and resilience ... I also believe they should be managed for the benefit of all equally, including ranchers/livestock owners.[19]

Canine Agency: The Wolf as More Than a Cipher

By way of conclusion, I want to suggest that a comprehensive understanding of the history and position of wolves (and indeed any animal) in North America is incomplete if it only entails a discussion of how they fit within the human world. A rising tide of scholarship is calling into question the idea of human exceptionalism and expanding our appreciation of the contribution of non-human animals and other beings to our world. At the same time much attention has been garnered by those arguing for the recognition of a new era of human primacy. While the competing literature regarding the so-called posthuman[20] and Anthropocene[21] eras continues to burgeon, there remains important middle ground to be carved out.[22] Alternative narratives need to be crafted that recognize animals as what they are apart from humans but that also acknowledge that, as humans, we cannot ever in fact regard them from outside of our humanity. This is truly one of the weightier conundrums of human consciousness.

While animals of all sorts can and are integrated into human cultural frameworks in ways that are multifaceted and complex, equally fascinating is the recognition that other animals also possess degrees of agency that, while never completely apart from the human meaning-giving, make them discernibly distinct in important ways. Thus, what I hope to do here, after a simple fashion, is not to discount the centrality of the human perspective but to lobby for an expansion of our appreciation of animals beyond the parameters of human meaning-giving. This endeavor has been labeled as "multispecies ethnography" and is "marked by its attentiveness to nonhuman agency – stones, plants, birds, and bees have the power to transform the world in this work."[23]

To initiate this final provocation to thought, it is probably best to describe what I mean by agency in this context. Agency is defined here as behavior whether of body or mind with a purpose; thus, intention is an integral part of understanding, from a human perspective, what constitutes agency.

One of the clearest ways to render a discussion of animal agency, perhaps, is to speak of how such agency inspires choices and provokes action by other animals, including humans. In the sections that follow I want to sketch an appreciation, based on ethnographic and statistical evidence from Montana, of how, within their "nature" as carnivores, wolves in the GYE exhibit agency to a degree that catalyzes a number of positive and negative responses by local human groups.

Two Ways of Being a Carnivore

Wolves are carnivores; as such they subsist on the flesh of other animals. Wolves in North America rely upon large ungulates as their primary prey. As predators of large ungulates, wolves have many choices in the GYE: in addition to relatively abundant wild species of moose, elk, bison, bighorn sheep, mountain goats, pronghorn and deer, wolves enjoy a veritable smorgasbord of domestic animals, such as goats, sheep and cattle, beginning just beyond YNP's intangible boundaries.

Regardless of which path wolves follow – subsisting on wild or domestic animals – they provoke both positive and negative reactions from local humans. If they kill and/or disrupt the lives, health and migration patterns of the trophy game species, especially elk, then wolves are considered "problems" by local hunters and professional outfitters and guides. The same practices are lauded by the members and staffs of local and national environmental organizations, scientists and public-lands administrators for the contribution to the re-creation of ecological stability in the GYE (more on this in the next section).

If they kill or terrorize domestic stock, then wolves are considered "problems" by local ranchers and farmers. This type of behavior results invariably in the wolves being killed. There are two curious things about this eventuality: (1) generally the wolves are not killed by the farmers and rancher but by agents of the US Fish and Wildlife Service (after the livestock depredation has been confirmed by a wildlife specialist); and (2) the livestock owner is reimbursed (by the Montana State Livestock Loss Board) for animals lost to predation.

The Montana State Livestock Loss Board was created by the state of Montana and charged with investigating claims of wolf and grizzly bear depredation on livestock and reimbursing the owners at market value for all confirmed kills. The board was created in 2009, when wolves were delisted as an endangered species following the approval of Montana's Wolf Conservation and Management Plan by the US Fish and Wildlife Service. At that time the not-for-profit organization Defenders of Wildlife, which had since 1995 provided reimbursement and non-lethal techniques for deterring wolf depredations, seeded the program with a $100,000 grant.

To provide a sense of its activities, here is a synopsis of the Montana State Livestock Loss Board numbers, as presented on its website (see Table 14.1). In 2014, which is the most recent year for which data is available, it paid out approximately $127,000 to 56 different owners for losses of livestock in Montana. Eight sheep, 63 cows, and two horses were confirmed as killed by predators, according to state wildlife agents. In addition, there were 15 "probable" kills, which are reimbursed at 50 percent of the market value. Of the total 88 head confirmed lost to predators, almost half (36) were killed by grizzlies. The total reimbursement for 2014 was the second highest yearly total in Montana since 2008, when the state began keeping statistics; it was also higher than the per-year average ($106,227) for that seven-year

Table 14.1 Livestock depredations, owners reimbursement, and wolves killed in Montana

	2008	*2009*	*2010*	*2011*	*2012*	*2013*	*2014*
Livestock Depredations	238	370	175	95	125	102	88
Reimbursement for Depredations ($US)	87,317	144,995	99,554	85,854	102,714	95,963	127,189
Wolves "Harvested"	N/A	72	N/A	166	175	231	N/A
Wolves "Controlled" by Officials/Citizens	N/A	145/11	163/16	64/19	108/31	75/26	N/A

Sources: Montana Official State Website, "Livestock Loss Statistics 2008–2014"; US Fish and Wildlife Services et al., *Rocky Mountain Wolf Recovery 2008 Interagency Annual Report*; US Fish and Wildlife Services et al., *Rocky Mountain Wolf Recovery Program 2009 Interagency Annual Report*; US Fish and Wildlife Services et al., *Rocky Mountain Wolf Recovery Program 2010 Interagency Annual Report*; US Fish and Wildlife Services et al., *Northern Rocky Mountain Wolf Recovery Program 2011 Interagency Annual Report*; US Fish and Wildlife Services et al., *Northern Rocky Mountain Wolf Recovery Program 2012 Interagency Annual Report*; US Fish and Wildlife Services et al., *Northern Rocky Mountain Wolf Recovery Program 2013 Interagency Annual Report.*

time frame. However, the 2014 total of combined, confirmed and probable, depredations of 88 was the lowest total for any year since such numbers were recorded.[24]

Plenty of Park County residents with whom I have spoken consider the livestock depredation reimbursement program a means to placate the powerful agricultural lobby in Montana. A participant in my research, "Samuel," called it a "patch of political theater," by which he means that it is program designed to undercut the potentially vociferous protests by farmers and ranchers if they had to shoulder the cost of the loss of livestock. "Ranchers can squawk pretty loud out here in Montana … their voices get heard, here [locally] and up in Helena [the state capital]." "Samuel's" point is that it is in the best interest of the pro-wolf people to keep the critique of Montana wolf management by the agricultural professional community, which is powerful economically, politically and culturally in the state, to a bare minimum.

Given the cultural climate of popular and political support for agriculture in Montana, wolves are quickly and easily cast in the role of marauding killer of domestic animals if they choose to prey upon livestock. If, by contrast, they subsist on wildlife, then it is easy for many Montanans to interpret and describe their behavior as being in line with their "natural" role. However, as we will see this too is seen as provocation by some members of the local communities.

Wolves as Agents of Greater Ecosystem Sustainability

In July 2013, the British environmental journalist George Monbiot made a presentation at the TEDGlobal 2013 Conference on the topic of "rewilding," the idea of humans returning animals, plants and processes to ecosystems as a means to reestablish the health and sustainability, ultimately, of the planet. In his talk, Monbiot discussed wolves in Yellowstone National Park as an important, prototypical example of this project and its potential. An outline of his point runs like this: wolves, as predators, chase ungulates, primarily elk [he says deer], which, in the absence of wolves, had increased in number beyond the GYE's carrying capacity. By pursuing elk as prey wolves contribute to lower numbers by killing them as well as by forcing them to disperse into smaller groups and chasing them away from the main riparian corridors to which they had grown accustomed. The less elk-related impact on these areas allowed grasses and hydrophilic trees, such as willows and cottonwoods, to return in abundance. Such plants and particularly their roots help secure the soil in such riparian areas. By stabilizing the banks, rivers with such plant life become less prone to flood, they develop more pools and meanders, which create slow, wide and irregular beds which, in turn, contribute to further plant regeneration as well as create greater habitat for fish, especially the native cutthroat trout, which benefit also from streams that run more clear of debris and slit. The point, as Monbiot makes it, is that the reintroduction of wolves has a "tropic cascade" effect across the ecosystem at a variety of levels, which has resulted in greater sustainability and a return to greater ecological stability.[25]

I mention this to illustrate the consequences of wolf reintroduction to the GYE, which has made it such a powerful example for ecologists both within and without the National Park Service. This group has applauded the agency of the wolves in dispersing the herds and reducing the numbers of elk in the Park.

However, as was noted earlier, wolf predation of wildlife in the GYE is not universally applauded by humans. In fact, there exists a very adamant segment of the local population of Park County that sees "wolves being wolves" as negatively impacting their recreational and professional patterns of life. In fact, wolf predation of elk, in particular, has become a significant issue of public discussion on the northern edge of the GYE.

First, some background: the vast elk population, estimated to be over 30,000 head, that summers in YNP breaks into two distinct herds in the fall when they migrate to avoid the harsh winters on the Yellowstone Plateau. The southern herd moves down into Jackson Hole in Wyoming, while the northern herd drifts into the lower elevations of southern Montana.

The northern herd has now for a century or more been the source for local hunters in Park County, Montana, both for meat and trophy animals. In the last several decades hunting by out-of-state residents, especially for trophy animals, has inspired the rise of a significant industry in professional guides and outfitters in the GYE of Montana, which is itself another great example of the creation of a post-industrial experience-based economy. It is these two groups – the local hunters and professional guides and outfitters – who are the loudest and most adamant opponents of wolf predation and terrorization of elk. The northern herd has seen a precipitous drop in its numbers since the 1990s, which is largely attributed to wolf predation and terrorization. Numbers that were estimated to be over 20,000 in the early 1990s dropped to an all-time low of 3,915 in 2013.

These low numbers prompted the Montana Fish, Wildlife and Parks to cancel the annual late season hunt in the Gardiner area of southern Park County, which had for decades been a staple of the local hunting opportunities. In addition, the Montana Fish, Wildlife and Parks curtailed elk hunting permits in the rest of southern Park County at that time as well. Thus, those invested in the experience economy that revolves around elk hunting, as producers (that is, guides and outfitters) or consumers (that is, hunters), are disgruntled.

To see an expression of this, one can look at the Rocky Mountain Elk Foundation, an organization catering to the interests of hunters as well as guiding and outfitting professionals. Since wolves that kill domestic species are considered "problem" animals by regulators, and almost invariably are killed as soon as possible, the agency of wolves has a lot to do with their success as a reintroduced species. If they choose to pursue and kill elk and bison, which are by far the most populace species in the YNP, then the wolves' agency has brought them in line with one of the largely unstated purposes of reintroduction: to reestablish an apical predator into the ecosystem, thereby culling and reducing the herds of elk and bison into, hopefully, sustainable populations. If they choose to pursue and kill domestic animals, wolves become prey themselves. Thus, wolves' agency is a key element in this equation that cannot be ignored.

Although this chapter stops short of serving as an example of the emerging genre of "multispecies ethnography," I feel that it is, in its own humble way, an example of a "project that seeks to understand the world as materially real, partially knowable, multicultured and multinatured, magical, and emergent through the contingent relations of multiple beings and entities."[26] Through this chapter I have also aspired to contribute to the posthumanist perspective, in that this work is inspired by a similar desire to "decenter the 'human' as a coherent, singular and Eurocentric subject, external to beings considered 'of nature,' such as other animals, and from 'naturalized humans,' such as indigenous peoples" as well as by a shared unease with "the consequences of human exceptionalism to other species and the environment."[27]

To my mind, an effective posthuman analysis should seek, for non-human entities, exactly what an effective post-colonial anthropology sought and seeks for non-Western subjects: to speak in ways that communicates the intention and nuance of the other-animal/colonized in ways that do not simply reflect the perspective of the human/colonizer but instead speaks in ways that the "other" would understand and recognize as their own. I think this project is a tall order – as Ogden, Hall and Tanita noted, "multispecies ethnography, at its heart, is a 'speculative' mode of inquiry that allows for speculative modes of writing"[28] – and this chapter is but one halting step in the direction of its inauguration.

Notes

1 Hines, "The Post-Industrial Regime," 74–97; Hines, "In Pursuit of Experience."
2 Weaver, "The Wolves of Yellowstone."
3 Cole, "Mission-Oriented Research."
4 McNaught, "Wolves in Yellowstone," 518–21.

5 The YNP wolf population had previously been composed of the subspecies of Northern Rocky Mountains wolf (*Canis lupus irremotus*). However, the introduced individuals were of the more common subspecies of Mackenzie Valley wolf (*Canis lupus occidentalis*).
6 Wyoming Game and Fish Department, "Annual Report."
7 I have conducted fieldwork annually in the GYE for over a decade beginning with a year of research in 2002 and another six months of full-time residence in 2014–2015; I have accumulated almost three years in the field, in which I have accomplished, in addition to general and focused observation, over 300 interviews with participants ranging from the directors of not-for-profit environment organizations, to a wide variety of local native citizens, to administrators within the US Forest Service and the US National Park Service.
8 Hines, "On Water and Wolves"; Wilson, "The Wolf in Yellowstone," 453–68.
9 The Wise Use movement was a "coalition of organizations … [whose] agenda concentrated on efforts by rural commodity producers to maintain their historical, privileged access to and control over the federally owned lands in the region." McCarthy, "First World Political Ecology," 1282.
10 Wilson, "Wolf in Yellowstone," 454.
11 Robson and Butler, "Coming to Terms with London," 71.
12 Liechty, *Suitably Modern*, 14.
13 Backus, "Bounty on Wolves, Lions."
14 Chaney, "Plan to Control County's Wolves."
15 Ibid.
16 Greater Yellowstone Coalition, "Wolves: Still Fighting for their Place in Greater Yellowstone."
17 Hines, "Rural Gentrification."
18 Kaufman, "New Dynamic in Wolf Country."
19 Defenders of Wildlife.
20 For example, Haraway, *The Companion Species Manifesto*; Haraway, *When Species Meet*.
21 For example, Steffen, Crutzen and McNeill. "The Anthropocene," 614–21; Kirksey and Helmreich, "The Emergence of Multispecies Ethnography," 545–76.
22 As Kirksey and Helmreich would have us appreciate, the Anthropocene and multispecies ethnography are not contradictory pursuits. Kirksey and Helmreich,"The Emergence of Multispecies Ethnography," 549.
23 Ogden, Hall and Tanita, "A Review of Multispecies Ethnography," 16.
24 Montana Official State Website, "Livestock Loss Statistics 2008–2014."
25 Monbiot, "Rewild the World."
26 Ogden, Hall and Tanita, "A Review of Multispecies Ethnography," 6.
27 Ibid., 7–8.
28 Ibid., 17.

Bibliography

Ethnographic Materials

All field journals, audio, and transcripts of the interviews are in the possession of the author.

Literature

Backus, P. "Livestock Producers Group Hopes to Put Bounty on Wolves, Lions." *Ravelli Republic*, February 22, 2012.

Chaney, R. "Activist Wants to Plan to Control County's Wolves." *Missoulian*, February 23, 2012.

Cole, G. F. "Mission-Oriented Research in the Natural Areas of the National Park Service." In *Research Note* 6. Mammoth, WY: Yellowstone National Park, 1969.

Defenders of Wildlife. Accessed February 17, 2012. www.defenders.org.

Greater Yellowstone Coalition. "Wolves: Still Fighting for Their Place in Greater Yellowstone." Accessed February 18, 2012, www.greateryellowstone.org/uploads/wolves_statementNEW_final.pdf.

Haraway, Donna. *The Companion Species Manifesto: Dogs, People, and Significant Otherness*. Chicago: Prickly Paradigm, 2003.

Haraway, Donna. *When Species Meet*. Minneapolis, MN: University of Minnesota Press, 2008.

Hines, J. Dwight. "In Pursuit of Experience: The Post-Industrial Gentrification of the Rural American West." *Ethnography* 11 (2010): 285–308.

Hines, J. Dwight. "Rural Gentrification as Permanent Tourism: The Creation of the 'New' West Archipelago as Post-Industrial Class-Cultural Space." *Environment and Planning D: Society & Space* 28 (2010): 509–25.

Hines, J. Dwight. "The Post-Industrial Regime of Production/Consumption and the Rural Gentrification of the 'New' West Archipelago." *Antipode* 44 (2012): 74–97.

Hines, J. Dwight. "On Water and Wolves: Toward an Integrative Analysis of the Rural Gentrification of the 'New' West." In *The Rural West: Toward a Regional Approach to Common Issues*, edited by David Danbom, 83–100. Salt Lake City, UT: University of Utah Press, 2015.

Kaufman, L. "After Years of Conflict, a New Dynamic in Wolf Country." *The New York Times*, November 4, 2011.

Kirksey, Eben, and Stefan Helmreich. "The Emergence of Multispecies Ethnography." *Cultural Anthropology* 25 (2010): 545–76.

Liechty, Mark. *Suitably Modern: Making Middle-Class Culture in a New Consumer Society*. Princeton, NJ: Princeton University Press, 2003.

McCarthy, James. "First World Political Ecology: Lessons from the Wise Use Movement." *Environment and Planning A* 34 (2002): 1281–302.

McNaught, D. "Wolves in Yellowstone Park? Park Visitors Respond." *Wildlife Social Bulletin* 15 (1987): 518–21.

Monbiot, George. "For More Wonder, Rewild the World." TEDGlobal 2013. Accessed May 18, 2015. www.ted.com/talks/george_monbiot_for_more_wonder_rewild_the_world.

Montana's Official State Website. "2008 Livestock Loss Statistics." Accessed November 20, 2015. http://liv.mt.gov/LLB/lossdata_2008.mcpx.

Montana's Official State Website. "2009 Livestock Loss Statistics." Accessed November 20, 2015. http://liv.mt.gov/LLB/lossdata_2009.mcpx.

Montana's Official State Website. "2010 Livestock Loss Statistics." Accessed November 20, 2015. http://liv.mt.gov/LLB/lossdata_2010.mcpx.

Montana's Official State Website. "2011 Livestock Loss Statistics." Accessed November 20, 2015. http://liv.mt.gov/LLB/lossdata_2011.mcpx.

Montana's Official State Website. "2012 Livestock Loss Statistics." Accessed November 20, 2015. http://liv.mt.gov/LLB/lossdata_2012.mcpx.

Montana's Official State Website. "2013 Livestock Loss Statistics." Accessed November 20, 2015. http://liv.mt.gov/LLB/lossdata_2013.mcpx.

Montana's Official State Website. "2014 Livestock Loss Statistics." Accessed November 20, 2015. http://liv.mt.gov/LLB/lossdata_2014.mcpx.

Ogden, Laura A., Billy Hall, and Kimiko Tanita. "Animals, Plants, People, and Things: A Review of Multispecies Ethnography." *Environment and Society* 4 (2013): 5–24.

Robson, G., and T. Butler. "Coming to Terms with London: Middle-Class Communities in a Global City." *International Journal of Urban and Regional Research* 25 (2001): 70–86.

Steffen, Will, Paul J. Crutzen, and John R. McNeill. "The Anthropocene: Are Humans Now Overwhelming the Great Forces of Nature?" *Ambio: A Journal of the Human Environment* 36 (2007): 614–21.

US Fish and Wildlife Service, Nez Perce Tribe, National Park Service, Montana Fish, Wildlife & Parks, Blackfeet Nation, Confederated Salish and Kootenai Tribes, Idaho Fish and Game, and USDA Wildlife Services. *Rocky Mountain Wolf Recovery 2008 Interagency Annual Report.* Accessed September 27, 2016. www.fws.gov/mountain-prairie/species/mammals/wolf/annualrpt08/FINAL_2008_Northern_Rockies_Summary_and_Background_3_17_09.pdf.

US Fish and Wildlife Service, Nez Perce Tribe, National Park Service, Montana Fish, Wildlife & Parks, Blackfeet Nation, Confederated Salish and Kootenai Tribes, Idaho Fish and Game, and USDA Wildlife Services. *Rocky Mountain Wolf Recovery 2009 Interagency Annual Report.* Accessed September 27, 2016. www.fws.gov/mountain-prairie/species/mammals/wolf/annualrpt09/FINAL_2009_Northern_Rockies_Summary_and_Background_3_3_10.pdf.

US Fish and Wildlife Service, Montana Fish, Wildlife & Parks, Nez Perce Tribe, National Park Service, Blackfeet Nation, Confederated Salish and Kootenai Tribes, Wind River Tribes, Washington Department of Wildlife, Oregon Department of Wildlife, Utah Department of Natural Resources, and USDA Wildlife Services. *Rocky Mountain Wolf Recovery 2010 Interagency Annual Report.* Accessed September 27, 2016. www.fws.gov/mountain-prairie/species/mammals/wolf/annualrpt10/FINAL_2010_Northern_Rockies_Summary_and_Background_3_9_11.pdf.

US Fish and Wildlife Service, Idaho Department of Fish and Game, Montana Fish, Wildlife & Parks, Nez Perce Tribe, National Park Service, Blackfeet Nation, Confederated Salish and Kootenai Tribes, Wind River Tribes, Washington Department of Fish and Wildlife, Oregon Department of Fish and Wildlife, Utah Department of Natural Resources, and USDA Wildlife Services. *Northern Rocky Mountain Wolf Recovery Program Interagency 2011 Annual Report.* Accessed September 27, 2016. www.fws.gov/mountain-prairie/species/mammals/wolf/annualrpt11/030612_final_nrm-background-summary_2011.pdf.

US Fish and Wildlife Service, Idaho Department of Fish and Game, Montana Fish, Wildlife & Parks, Nez Perce Tribe, National Park Service, Blackfeet Nation, Confederated Salish and Kootenai Tribes, Wind River Tribes, Confederated Colville Tribes, Washington Department of Fish and Wildlife, Oregon Department of Fish and Wildlife, Utah Department of Natural Resources, and USDA Wildlife Services. *Northern Rocky Mountain Wolf Recovery Program Interagency 2012 Annual Report.* Accessed September 27, 2016. www.fws.gov/mountain-prairie/species/mammals/wolf/annualrpt12/FINAL_NRM-Summary_2012.pdf.

US Fish and Wildlife Service, Idaho Department of Fish and Game, Montana Fish, Wildlife & Parks, Wyoming Game and Fish Department, Nez Perce Tribe, National Park Service, Blackfeet Nation, Confederated Salish and Kootenai Tribes, Wind River Tribes, Confederated Colville Tribes, Spokane Tribe of Indians, Washington Department of Fish and Wildlife, Oregon Department of Fish and Wildlife, Utah Department of Natural Resources, and USDA Wildlife Services. *Northern Rocky Mountain Wolf Recovery Program Interagency 2013 Annual Report.* Accessed September 27, 2016. www.fws.gov/mountain-prairie/species/mammals/wolf/annualrpt13/reports/FINAL_NRM-Sum2_2013.pdf.

Weaver, J. L. "The Wolves of Yellowstone." In *National Park Service National Resources Report* 14. Washington, DC: Government Publishing Office, 1978.

Wilson, M. "The Wolf in Yellowstone: Science, Symbol, or Politics? Deconstructing the Conflict between Environmentalism and Wise Use." *Society & Natural Resources* 10 (1997): 453–68.

Wyoming Game and Fish Department, 2008. "Annual Report." Accessed September 27, 2016. https://wgfd.wyo.gov/About-Us/Game-and-Fish-Commission/Game-and-Fish-Commission-Annual-Reports.

Epilogue

Harriet Ritvo

Animal agency has preoccupied the field of animal studies since its emergence in the last decades of the twentieth century. Often this preoccupation has had a political or argumentative edge. Asserting that non-human animals have agency is a way of asserting that they have standing – that they matter – that they deserve consideration. The proximate audience for most such assertions consists of like-minded humanists and social scientists (almost all scholars who choose to study human–animal relations agree that their subjects merit consideration, sympathy, and respect). But the ideal audience is much broader, since even a cursory survey of the experience of most animals at human hands or in human power will demonstrate that this attitude is far from universal. The presumably happy lives of cherished companion animals, numerous though they are, unfortunately remain the exception rather than the rule.

If there is consensus within animal studies about the existence of animal agency, agreement about its exact nature remains elusive. The difficulty of arriving at a clear, concise, and widely acceptable definition is illustrated by several of the essays collected in this volume. They offer surveys of the scholarly debate that are similarly detailed and erudite, but not otherwise identical. They approach the issue from different perspectives and cite different (although overlapping) lists of authorities. In part this divergence reflects their varied disciplinary backgrounds. Because animal studies is a field, rather than a discipline, the work of scholars who situate themselves within it exemplifies a wide range of methodologies and theoretical understandings. And in part this divergence reflects the fact that defining animal agency depends on the resolution of a broader question: what is agency more generally? This question continues to engage philosophers, and so lack of consensus among historians, sociologists, anthropologists, political ecologists, and cultural studies scholars may not be surprising.

This volume also shows, however, that even in the absence of agreement in the abstract, concrete practices may converge. The essays in this volume reveal a shared sense that animals have intentions and act on them. (It should be noted that this shared sense implies a restricted definition of "animal," a term that is subject to at least as much interpretation and debate as "agency." Thus each of the contributors focuses on one or several individual species, but none of those species are invertebrate.) And whether they are investigating the agency of horses or dogs or herrings or rabbits or wolves, they also tend to assume that it somehow involves cooperation with humans. Sometimes this cooperation is explicit and obvious to both parties, as is the case with such domesticated animals as horses and dogs. Sometimes animals enact their agency by violating these mutual understandings. And sometimes the collaboration is involuntary and unconscious, as with the contribution of white-tailed eagles to the awakening of environmental consciousness or the shared work of fishermen and herring in the creation of a socio-ecological community in the Baltic. *Shared Lives of Humans and Animals* thus demonstrates how illuminating it is to acknowledge and assume the agency of other animals, whatever that agency turns out to be.

Index

For Product Safety Concerns and Information please contact our
EU representative GPSR@taylorandfrancis.com Taylor & Francis
Verlag GmbH, Kaufingerstraße 24, 80331 München, Germany